A METHODOLOGY FOR DEVELOPING ARMY ACQUISITION STRATEGIES FOR AN UNCERTAIN FUTURE

John E. Peters
Bruce Held
Michael V. Hynes
Brian Nichiporuk
Christopher Hanks
Jordan Fischbach

Prepared for the United States Army

RAND ARROYO CENTER

The research described in this report was sponsored by the United States
Army under Contract No. W74V8H-06-C-0001.

Library of Congress Cataloging-in-Publication Data

A methodology for developing Army acquisition strategies for an uncertain future /
John E. Peters ... [et al.].
    p. cm.
    Includes bibliographical references.
    ISBN 978-0-8330-4048-0 (pbk.)
    1. United States. Army—Procurement. I. Peters, John E., 1947–

UC263.M44 2007
355.6'2120973—dc22

                        2007009908

Published 2007 by the RAND Corporation
1776 Main Street, P.O. Box 2138, Santa Monica, CA 90407-2138
1200 South Hayes Street, Arlington, VA 22202-5050
4570 Fifth Avenue, Suite 600, Pittsburgh, PA 15213-2665
RAND URL: http://www.rand.org/
To order RAND documents or to obtain additional information, contact
Distribution Services: Telephone: (310) 451-7002;
Fax: (310) 451-6915; Email: order@rand.org

Preface

Army acquisition investment strategy confronts many challenges, among them, responding rapidly to the evolving needs of its soldiers deployed in the war on terrorism, supporting ongoing efforts to transform the force and develop the Future Combat System (FCS), and maintaining enough flexibility and responsiveness to meet the Army's needs when confronting unanticipated circumstances and adversaries around the world. This monograph develops a methodology to adjust the Army's acquisition investment strategy within current regulatory guidelines but in ways that could yield greater flexibility and responsiveness to the needs of the service.

The research was sponsored by the office of the Assistant Secretary of the Army for Acquisition, Logistics, and Technology and was conducted within RAND Arroyo Center's Force Development and Technology program. RAND Arroyo Center, part of the RAND Corporation, is a federally funded research and development center sponsored by the United States Army.

The monograph should be of interest to those concerned with acquisition policy and practices and to a broader audience interested in Army modernization and transformation.

For more information on RAND Arroyo Center, contact the Director of Operations (telephone 310-393-0411, extension 6419; fax 310-451-6952; email Marcy_Agmon@rand.org) or visit Arroyo's web site at http//www.rand.org/ard/.

Contents

Figures

Tables

Summary

The Army acquisition community stands at a critical juncture. The Future Combat System, the centerpiece of Army transformation, has proven to be more expensive and technologically more complicated than originally anticipated, and the rapid pace of ongoing operations means that many key weapon systems will reach the end of their service lives sooner than planned or will require intensive maintenance to keep functioning. The future presents even more challenges for which the Army must prepare, including a wide range of dangerous adversaries, the potential reallocation of combat tasks across and among the services, and the prospect of budget pressures.

Taken together, these circumstances raise some important questions for the Army acquisition community. In particular, what should a robust acquisition investment strategy look like—one designed to perform well against all of the anticipated threats? Further, how should the Army acquisition community assess the appropriateness of its investment strategy as time goes by? This study seeks to provide insight into these questions by describing a new way for the Army to assess investments across a broad range of options. This method, the Acquisition Investment Management (AIM) model, incorporates Assumption-Based Planning (ABP), a tool developed by RAND to assist in planning during uncertain times.[1]

---

[1] Dewar (2002).

Assumption-Based Planning Can Be Used to Assess Army Acquisition Plans

ABP is a technique for evaluating plans to ascertain the degree to which they rely on assumptions that might be vulnerable. We used this technique to assess the Army acquisition community's current plans and to determine whether they were robust or resting on fragile, questionable assumptions. ABP is based on the notion that an organization's operations or plans will change if its corresponding underlying assumptions about the world change. The main steps in the ABP process are shown in Figure S.1.

We applied the five main steps in the ABP process to Army acquisition policy, first by identifying the assumptions that underlie that policy. Next, we identified *load-bearing assumptions*, i.e., important assumptions that underpin and shape Army acquisition plans. If a load-bearing assumption fails or becomes "broken," the organization's plans would be at risk. Therefore, we identified a series of *signposts*, i.e.,

Figure S.1
The Assumption-Based Planning Process

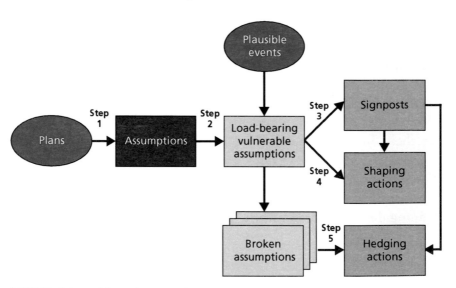

SOURCE: Adapted from Figure 3.1 in Dewar (2002).
RAND *MG532-S.1*

indicators that an assumption is becoming vulnerable. We also identified *shaping actions* that can be used to keep assumptions viable and *hedging actions* that can be taken to prepare for unwelcome but unpreventable developments. For the acquisition community, shaping and hedging actions both take the form of investments.

The AIM Methodology Also Incorporates Information About Current and Potential Future Threat Environments and Likely Army Involvement

The traditional ABP process would involve just the five steps described in the previous paragraph. However, AIM goes further by also incorporating information about current and alternative threats and the relative likelihood of Army involvement in each. This step was needed to ensure that the Army's acquisition strategy can respond to emerging new threats that require high levels of Army involvement. To understand the threat environment, we examined the current Department of Defense (DoD) *Strategic Planning Guidance* and, more specifically, its threat characterization of irregular and conventional adversaries and of disruptive to catastrophic effects.[2] We evaluated each threat in terms of its level of potential consequences for the United States and the potential likelihood of Army involvement. We used this information to build a plot that relates the level of threat posed to the United States (disruptive to catastrophic) to the likelihood and depth of Army involvement (from low to high). We populated the plot with alternative sets of plausible circumstances, as shown in Figure S.2.

The AIM Methodology Can Be Used to Specify an Army Investment Strategy

The ultimate objective of the AIM process is to identify an appropriate balance of investments that takes into account the relative severity of

---

[2] U.S. Department of Defense (2004b).

Figure S.2
Threats and Likelihood of Army Involvement

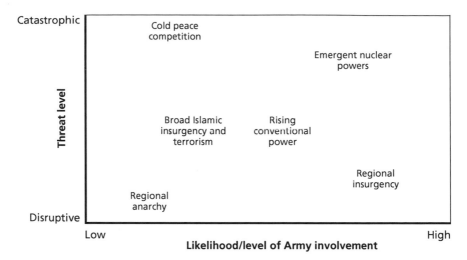

the plausible alternative circumstances and the relative likelihood of Army involvement. To do this, we associated each signpost of vulnerability with different alternative circumstances and related shaping and hedging actions. For the acquisition community, the appropriate responses to signposts are investments: investments in shaping activities to negate the effect of dangerous signposts and investments in hedging activities to cope with circumstances when an assumption becomes vulnerable and begins to fail.

Thus, as the alternative sets of circumstances move about the plot of threat and likelihood over time—as measured by intelligence products and the collective judgment of the leadership—this movement engages the shaping and hedging strategies, expressed in terms of adjustments to acquisition investments. The resulting redistribution of funds across accounts becomes the adjusted acquisition investment strategy. The strategy is biased toward circumstances that seem more likely and most dangerous and thus gives the Army the flexibility, agility, and responsiveness to meet emerging conditions.

The AIM Methodology Can Be Used at Several Points in the Army's Planning and Programming Process

If adopted, the AIM methodology would allow the acquisition community to review its plans and investment decisions regularly to check their congruence with national-level planning guidance. The process would also provide a check on the recommendations from the bottom-up activities that typify current acquisition and force development decision support. Thus, the new process overcomes the problem of competing visions and preferences and replaces them with a new calculus that acquisition leaders can use to guide their investment strategy decisionmaking.

Because AIM is designed to help increase the *strategic* responsiveness of materiel acquisition in the Army, it should be used as part of the planning and programming—not the budgeting—process. AIM might be incorporated into existing processes and activities in several ways.

The Office of the Assistant Secretary of the Army for Acquisition, Logistics, and Technology (OASA(ALT)) could use AIM to help support G-8 in maintaining and updating the Research, Development, and Acquisition Plan (RDAP) database for the Program Objective Memorandum (POM). The RDAP is a forward-looking 15-year plan that provides a detailed view of what the Army intends to spend on the development and production of technologies and materiel. These RDAP updates could be used to help inform the development of Army POMs, which specify how the Army intends to allocate its budget. Under the current biennial approach, "full" POMs, i.e., POMs that address how every dollar is allocated, are prepared only in even calendar years. In odd calendar years, the full POM from the preceding year may be updated as necessary. Therefore, OASA(ALT) would perform AIM runs in time to support the regularly scheduled RDAP updates that G-8 makes each year, whether before a "full" POM-build in an even year or a "POM-update" in an odd year.

The G-3 could use AIM outputs to help support the Army's input to the DoD Strategic Planning Guidance (SPG). G-3 is responsible for providing strategic planning guidance to the Army planning and programming process; this guidance is used by the Secretary of Defense to drive overall DoD planning and programming. This would be a natural place to incorporate AIM outputs.

The Army's deliberative forums for assembling the POM could also use AIM in performing their development, oversight, and review roles. These forums include the Council of Colonels, the Planning Program Budget Committee (PPBC), the Senior Review Group (SRG), and the Army Resources Board. All of these have a role in the POM-building process.

AIM outputs could also play a role in the capabilities needs analysis (CNA) that Training and Doctrine Command (TRADOC) performs to assess battlefield materiel capabilities and determine modernization alternatives for input to the Army POM. The CNA is an interactive process involving TRADOC's schools, proponents, and Army Headquarters staff. AIM could help inform those interactions and the conclusions they produce regarding how materiel-development resources should perhaps be reallocated if important changes occur in the projected national security environment.

Recommendations for Implementing AIM

The AIM process relies on judgments about threats and likelihood and on acquisition officials to interpret intelligence reports and make decisions—decisions some officials may believe are beyond their authority or that are best the product of group judgment and consensus. The process may also seem to require the inputs of subject matter experts, many of whom may not reside in the acquisition community. Such concerns can be properly addressed through carefully coordinated staff actions, such as the following:

- Form an acquisition strategy working group. An informal (although official) working group could be formed to meet periodically to consider the location of alternative sets of circumstances within the threat-likelihood plot and to conduct the necessary assessments and recommend reallocations of funds across accounts in response to movement of some of those conditions. The working group might include G-2 staff members and other intelligence officers, representatives from the Combatant Commands, congressional liaison, and the Joint Capabilities Integration Development System (JCIDS) community. It would also be prudent to include representatives from the Program Executive Office (PEO), and perhaps a representative from Office of the Secretary of Defense (OSD) Industrial Policy.

- **Consider other alternative circumstances.** The different sets of circumstances posited in this monograph represent our translation of global security trends onto the Strategic Planning Guidance's threat-likelihood space. Other analysts might have better information and might populate that space somewhat differently. Moreover, as time passes, other concerns may become plausible. Therefore, the Army acquisition community should periodically convene the acquisition strategy working group to consider new influential factors that could recast the threat-likelihood space. Such meetings might include the relevant National and Defense Intelligence Officers responsible for the regions and topics of concern.

- **Exploit Assumption-Based Planning.** It is important to review key plans periodically and to search them for indications of new or different assumptions. Where new assumptions are found, the acquisition community will want to generate new signposts and associated shaping and hedging strategies. The signposts and shaping and hedging should then be associated with alternative sets of circumstances to keep the acquisition investment strategy development process current.

- **Plan acquisition investment strategy reviews.** We also recommend that the acquisition community establish a schedule to begin these activities. The process might begin in the off-budget year. One possible approach would be to look for emergent alternative futures and to consider where alternative futures might lie in the threat-likelihood plot every other year.

Acknowledgments

We thank MG Jeffrey A. Sorenson, Deputy for Acquisition and Systems Management, within the Office of the Assistant Secretary of the Army for Acquisition, Logistics, and Technology, for his sponsorship of our work. We are also indebted to certain members of General Sorenson's staff, especially to LTC Michelle Nasser, who served ably as our project monitor, and to Kathy Finigan and Kenneth Murphy for their help and support. We thank Thomas McNaugher and Tim Bonds at RAND for their thoughtful advice and direction and Holly Johnson for her ever-skillful work in preparing the manuscript for publication.

Acronyms

ABP	Assumption-Based Planning
ACEIT	Automated Cost Estimating Integrated Tool
ACR	Armored Cavalry Regiment
ACTD	Advanced Concept Technology Demonstration
AIM	Acquisition Investment Management
ALT	Acquisitions, Logistics, and Technology
ARB	Army Resources Board
ASA(ALT)	Assistant Secretary of the Army for Acquisition, Logistics, and Technology
ASA(FM&C)	Assistant Secretary of the Army for Financial Management & Comptroller
ASPG	Army Strategic Planning Guidance
BES	Budget Estimate Submission
BOS	Budget Operating System
C4ISR	command, control, communications, computers, intelligence, surveillance, and reconnaissance
CBP	capabilities-based planning

CIA	Central Intelligence Agency
CMOC	Civil-Military Operations Center
CNA	capabilities needs analysis
COCOM	Combatant Command/Commander
CONOPS	Concept of Operations
CONUS	continental United States
COSCOM	Corps Support Command
COTS	Commercial Off the Shelf
CS	Combat Support
CSS	Combat Service Support
DAB	Director of the Army Budget
DAG	Defense Acquisition Guidebook
DCS	Deputy Chief of Staff
DIA	Defense Intelligence Agency
DISCOM	Division Support Command
DoD	Department of Defense
DPAE	Director, Program Analysis and Evaluation
EE PEG	Equipment Program Evaluation Group
EPP	Extended Planning Period
FCS	Future Combat System
FM&C	Financial Management and Comptroller
FYDP	Future Years Defense Plan
HNS	host nation support
ISR	intelligence, surveillance, and reconnaissance

JCIDS	Joint Capabilities Integration Development System
JSCP	Joint Strategic Capabilities Plan
JTRS	Joint Tactical Radio System
MDEP	Management Decision Package
MOS	military occupational speciality
MTOE	modified table of organization
NIE	national intelligence estimate
NTC	National Training Center
O&M	operations and maintenance
OASA	Office of the Assistant Secretary of the Army
OPFOR	opposing force
OPTEMPO	operations tempo
ORD	Operational Requirements Document
OSD	Office of the Secretary of Defense
PAE	Program Analysis and Evaluation
PEG	Program Evaluation Group
PEO	Program Executive Office
POM	Program Objective Memorandum
PPBC	Planning Program Budget Committee
PPBE	planning, programming, budgeting, and execution
PPBS	Planning, Programming, and Budgeting System
QDR	Quadrennial Defense Review
R&D	research and development
RDA	research, development, and acquisition

RDAP	Research, Development, and Acquisition Plan
RDT&E	research, development, test, and evaluation
RO/RO	roll-on/roll-off
ROTC	Reserve Officers' Training Corps
RSTA	reconnaissance, surveillance, and target acquisition
SIPRI	Stockholm International Peace Research Institute
SOF	special operations forces
SPG	Strategic Planning Guidance
SRG	Senior Review Group
T&L	threat and likelihood/level of Army involvement
TDA	table of distribution and allowances
TOA	Total Obligation Authority
TRADOC	Training and Doctrine Command
TRL	technology readiness level
VCSA	Vice Chief of Staff, Army
WARBUCS	Web Army RDA Budget Update Computer System
WMD	weapons of mass destruction

CHAPTER ONE

Introduction

Post-Conflict Acquisition Planning

Paradoxically, for most armies the period following the successful conclusion of a major conflict is often a difficult time of institutional stress. Demobilization, seriously reduced budgets, and uncertainty about future threats wreak havoc in even the best armies. In some instances, legislatures eye army budgets as potential bill-payers to sustain national entitlement programs in such areas as public health, education, and child welfare. In addition, major conflicts typically provoke enormous technological advances that offer opportunities for the army that is able to make the necessary doctrinal, materiel, and human investments required to harvest the new capabilities. These new opportunities, however—particularly those that arrive late in the conflict—often remain untried because of constrained resources.

The U.S. Army is no exception. In the aftermath of World Wars I and II and Vietnam, the U.S. Army was downsized and highly resource-constrained. Each of these post-conflict periods also witnessed debate about the effect that new technologies would have on how the U.S. Army fights. Over time, however, each post-conflict period ended and the Army adjusted. The U.S. Army went through a similar experience at the end of the Cold War. The Clinton administration cut budgets, especially in procurement, while at the same time Army forces were deployed in a host of smaller crises and operations other than war. With the advent of the Bush administration and the trauma of September 11, 2001, the Army entered another new era, one with arguably even more profound uncertainties that add complexity to Army planning. How the Army adjusts to its evolving security environment will determine how effective it will remain.

Among an Army's functions perhaps hardest hit by post-conflict pressures and uncertainties are the development and acquisition of new weapon systems and equipment. The combination of strategic uncertainty, promising but untried and immature technologies, reduced budgets, large wartime stocks of serviceable materiel, and expectations of prudent, even frugal, budget decisions all combine to make planning and implementing an effective acquisition strategy difficult at best. Because of the disruption and stress it has experienced, an Army acquisition system emerging from a post-conflict period needs a way to systemically plan its investments to develop and provide the materiel required in the new security environment. This seems particularly true today, as the U.S. Army acquisition system continues to grapple with the consequences of the 1990s procurement holiday and the many small-scale deployments and crises of that decade.

Since the end of that period, planning, programming, and budgeting the Army's materiel acquisitions have grown more difficult. Perhaps most significantly, the adversaries and the missions that the Army must be prepared for are more ambiguous and diverse than at any time since the period between the World Wars. Additionally, the pace of technological advance creates opportunities for the Army to transform, but it also presents a number of challenges, including preventing technical surprise, managing a transformation while highly resource-constrained, and developing concepts and doctrine for the use of new technologies. To complicate matters further, the pace of operations has been such that many key weapon systems will reach the end of their service lives sooner than planned, and some will require intensive maintenance to keep them functioning. Finally, the acquisition community is concerned that, as joint warfighting continues its evolution and as different battlefield tasks migrate to different elements within the joint force, it becomes more difficult to anticipate where Army acquisition investments will be most needed, because some traditional Army functions (e.g., fire support) may be provided by air and naval forces more frequently than in the past. In sum, the Army's challenge is to manage a resource-constrained acquisition strategy that will nevertheless be expected to field systems with capabilities suitable for operations against a wide range of adversaries under widely varying circumstances.

Taken together, these circumstances raise the question of whether the process used to plan, program, and budget the Army's acquisition activities is adequate for the task. On the whole, the answer appears to be that it is. The process is a sophisticated one that has been honed for more than 50 years. It is directed by top-down strategic guidance and informed by bottom-up statements of warfighting capability needs. Nevertheless, the core Army processes for planning, programming, and budgeting were formed in an era of superpower competition.[1] Since the boundaries of the strategic guidance were based on a national military strategy focused on the Soviet Union, the strategic guidance, although never simple, could be made understandable enough to carry through the Planning, Programming, and Budgeting System (PPBS).[2] Within this system, shaping programs and budgets in accordance with the guidance was relatively straightforward, so the acquisition program and budget were developed in an essentially competitive manner, with program advocates arguing the efficacy and alignment of particular programs within the context of the strategic guidance.

Today's emerging complex threat and mission environment suggests that aligning the Army acquisition program and budget with the strategic guidance is no longer so straightforward; planners may have a difficult time identifying choices that produce the closest fit between their planning and programming and the planning guidance. This implies that the competitive "horse-trading" that currently helps shape the overall acquisition program budget at the level of individual programs would benefit from a methodology to manage the tradeoffs against the strategic guidance. What we propose in this monograph, therefore, is the development of a planning tool to help the Army's programmers and budget personnel maintain the alignment between

---

[1] These processes are outlined in Chapter Three.

[2] On the origins of PPBS, see Enthoven (2005).

the strategic guidance and a developing Army acquisition program and budget.

As suggested above, alignment of the strategic guidance and the Army's acquisition program has become more difficult in recent years because there is more uncertainty concerning (1) the nature of future foes, (2) the types of conflict the Army must be prepared for, and (3) the tools and technologies available to the Army. Therefore, any planning assistance tool must help to

- plan a program to perform well against all of the threats identified in the Department of Defense (DoD) *Strategic Planning Guidance*
- create an investment strategy that is flexible enough to adapt to misunderstandings of the national security environment and, more important, to rapidly changing circumstances.

The dual requirements—to perform against a diverse mission set and to maintain responsiveness to a changing security environment— suggest methodologies associated with portfolio management and optimization strategies. However, given the uncertainty associated with estimating current and future security environments and the competing goals of the Army's acquisition program, these methodologies did not seem fully appropriate. Instead, this study describes and illustrates an adaptation of Assumption-Based Planning (ABP), a tool developed by RAND to assist in planning during uncertain times.[3] This study uses ABP to create an Acquisition Investment Management (AIM) model, which recommends acquisition investments across a broad range of capabilities. AIM contains both elements of portfolio management and a flavor of optimization. Instead of the maximization or optimization goals of those methodologies, however, AIM works toward a more realistic goal of satisfying the complex and evolving requirements in the national security guidance. In the remainder of this introduction, we describe ABP and explain our approach for applying it to the acquisition process.

---

[3] Dewar (2002).

Background on Assumption-Based Planning

Assumption-Based Planning is based on the notion that an organization's operations or plans will change if its corresponding underlying assumptions about the world change. The ABP process involves a series of questions that are used to examine an organization's plans and planning documents:

- What are the key, load-bearing assumptions underlying the plan?
- What signposts, if they appeared, would signal that these key assumptions were failing?
- What actions might be taken to shape the future environment to prevent these key, load-bearing assumptions from failing?
- In the event that the shaping strategy fails, what hedging steps can be taken to help cope with the emergent circumstances?

Key components of the ABP process include load-bearing assumptions, signposts, shaping strategies, and hedging actions. A *load-bearing assumption* is an important assumption that underpins and shapes an organization's plans. If a load-bearing assumption fails or becomes "broken," the organization's plans would be at risk. The ABP process therefore identifies and monitors the emergence of *signposts*, i.e., indicators that an assumption is becoming vulnerable. When a key assumption becomes vulnerable, the ABP process seeks to identify *shaping* actions that can be used to keep the assumption viable. In some cases, however, circumstances do not allow shaping; the ABP process then identifies *hedging actions*—steps that can be taken to prepare the organization for unwelcome but unpreventable developments.

Figure 1.1 outlines the Assumption-Based Planning process. A detailed description of ABP as well as of the Army's key load-bearing assumptions and their vulnerabilities can be found in Appendix A.

Figure 1.1
The Assumption-Based Planning Process

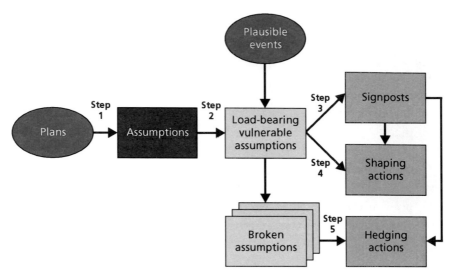

SOURCE: Adapted from Figure 3.1 in Dewar (2002).
RAND *MG532-1.1*

Research Approach

Now that we have defined the ABP process, we can explain how this process was applied as part of our overall research approach.

Overview

Our analysis had two main parts:

- First, we used ABP as part of a process the Army can use to build an acquisition investment strategy that is both robust and adaptive and thus able to cope with new developments as they unfold.
- Second, we applied this acquisition investment strategy development process to provide a "strawman" that the acquisition community can follow. This discussion includes the steps necessary to

integrate the strategy with the Army's programming and budgeting activities.

To provide additional perspective on the current situation facing the Army's acquisition community, we also examined a set of analogous circumstances from the interwar years (1919–1939).

We are confident that our application of the investment strategy formulation process is sound, but we also recognize that none of us is a member of the acquisition community and we do not possess the same skills and experiences as members of that community. Therefore, it is important that acquisition professionals implement the strategy development process for themselves. If the process described in this monograph is adopted, it will provide the acquisition community with a way to confirm the results of the bottom-up analysis of Army acquisition investment needs and a way to bring long-term coherence to the overall effort through periodic audits.

Research Steps

Identifying Key Assumptions and Vulnerabilities as Well as Alternative Threat Environments.

We began by identifying key assumptions that underlie the Army's acquisition strategy and plausible sets of circumstances that could significantly affect the Army's plans. There were two aspects to this process. First, we mined Army planning documents to find the key, load-bearing assumptions that serve as the foundation for the Army's acquisition strategy. These assumptions were used as the basis for developing signposts and, subsequently, shaping and hedging actions. Second, we identified features of the current and alternative threat environments and the relative likelihood of Army involvement in each. This latter step was necessary because the Army's acquisition strategy also needs to be able to respond to the emergence of alternative threats. A full description of the alternative sets of circumstances identified is found in Appendix B.

To understand the threat environment, we examined the current DoD *Strategic Planning Guidance* and, more specifically, its threat characterization of irregular and conventional adversaries and of disruptive to catastrophic effects.[4] We placed each dimension of the threat along a continuum. In our conceptualization, future adversaries existed along a scale ranging from disruptive to catastrophic. These adversaries ranged in type from traditional foes including Iran and Syria to less conventional, asymmetric enemies, such as North Korea and al Qaeda. In a similar fashion, we placed potential conflicts along a continuum based on their level of potential consequences: At the low end are conflicts leading to disruptive consequences, such as long lines at the filling station;[5] at the high end are conflicts with catastrophic consequences that could threaten the survival of the Republic, such as multiple, high-yield nuclear detonations in America's largest cities. Assessments of Army involvement are based on (1) the Army's historical engagement in the theater of operations, and (2) our assessment of the utility of Army capabilities for addressing the threat.

Next, we built a plot (shown in Figure 1.2) that relates the level of threat posed to the United States (disruptive to catastrophic) to the likelihood and depth of Army involvement (from low to high), if such events came to pass.[6] We populated the plot with alternative sets of plausible circumstances drawn from the potential conflicts identified in the previous step.[7] These alternative circumstances illustrate the basic concept emphasized in the *Strategic Planning Guidance,* that tomorrow might hold a large number of challenges, some of them conventional, others irregular, and ranging in their destructive potential from disruptive to catastrophic.

---

[4]   U.S. Department of Defense (2004b).

[5]   We have expanded on the DoD conception of "disruptive" threats, which much of the department's literature tends to treat as high-technology-based. We agree that high technology used against the United States might be disruptive, but we also believe that other threats, arguably those imperfectly executed, might have disruptive results as well.

[6]   Likelihood and level of Army involvement are combined into one metric to assess how involved the Army is likely to be. Although likelihood is a critical parameter, the Army will probably be involved at a low level in many things, so likelihood by itself is not very useful. This is best illustrated through example. Use of Army assets to combat the narcotics trade is highly likely (e.g., training military forces in Latin America) but should probably count for very little because the total percentage of Army resources involved will be very small.

[7]   Descriptions of the alternative sets of circumstances appear in Appendix B.

Figure 1.2
Threats and Likelihood of Army Involvement

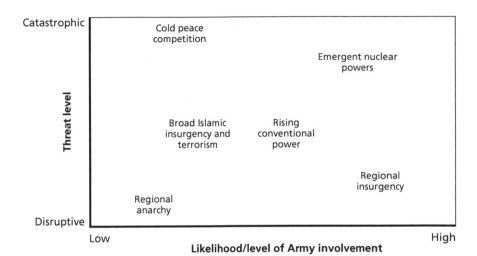

Using Signposts, Shaping Actions, and Hedging Actions to Identify Appropriate Investments. We next tested the fundamental assumptions underpinning the Army's acquisition system to determine whether they might be vulnerable: That is, we identified signposts of change in the security environment that could threaten the core assumptions and possibly render them "broken."[8] We associated these signposts with the alternative sets of circumstances identified in the previous step. The association between circumstances and signposts provides a way to track the signposts, i.e., as intelligence estimates and the professional judgment of acquisition officials lead us to believe that a given set of circumstances (or something closely resembling it) is emerging, the signposts signaling the vulnerability of Army assumptions will also emerge, and the acquisition community can take appropriate shaping and hedging actions in response. For the acquisition

---

[8] For example, presume that a fundamental assumption is that "the Army will maintain technical dominance over potential foes." A signpost of vulnerability would be if a potential foe began to significantly increase defense Research, Development, Test, and Evaluation (RDT&E) investments.

community, the appropriate responses to signposts are investments: investments in shaping activities to negate the effect of dangerous signposts, and investments in hedging activities to cope with circumstances when an assumption becomes vulnerable and begins to fail.

Thus, as the alternative sets of circumstances move about the plot of threat and likelihood over time—as measured by intelligence products and the collective judgment of the leadership—this movement engages the shaping and hedging strategies, expressed in terms of adjustments to acquisition investments. The resulting redistribution of funds across accounts becomes the adjusted acquisition investment strategy.

Specifying an Army Investment Strategy. The final research step was to design a specific Army investment strategy that would perform well when confronting any of the circumstances presented in the threat versus likelihood plot. The strategy is biased toward circumstances that seem more likely and most dangerous (i.e., the alternative circumstances in Figure 1.2) and that give the Army the flexibility, agility, and responsiveness to meet emerging conditions. A list of budget categories used in this analysis is found in Appendix C. The new, robust investment strategy that emerges from this research effort will not last indefinitely. The acquisition community must periodically monitor its health and appropriateness for the circumstances of the day. This monograph offers an investment review process to guide such an effort.

Insights from Recent History

In addition to developing and applying a new acquisition investment strategy, we also considered the way the Army has managed its acquisition accounts historically and the role of acquisition in Army management over time. This information is instructive for understanding more contemporary Army acquisition actions and decisions. Of particular interest are the difficulties experienced by the Army in the years between the two World Wars (1919–1939)—a time when the Army confronted no obvious significant threat. Until the early 1930s, Germany was disarmed, the League of Nations and the Kellogg-Briand Pact were expected by many to manage international security, and consensus within the U.S. Army was that the greatest threat to American security was on the border with Mexico. These issues are explored briefly in Chapter Four and in more detail in Appendix D.

Organization of This Monograph

The remainder of this monograph consists of three chapters. Chapter Two describes how we developed and applied the Army acquisition investment strategy process. Chapter Three discusses how this process could be incorporated into the Army's current programming and budgeting activities. Chapter Four concludes with lessons for the acquisition community drawn from the period between the two World Wars. Four appendixes support the analysis with details of Assumption-Based Planning, alternative sets of circumstances, the budget categories the research employed to create the acquisition investment strategy development process, and an account of the interwar era.

CHAPTER TWO

Developing a Tool for Investment Strategy Planning

In this chapter, we provide a detailed description of the methodology used in this study. We begin with an overview of the methodology as a whole and then describe each step individually. Our method draws on a capabilities-based planning (CBP) framework. The CBP system was started at the time of the 2001 *Quadrennial Defense Review* (QDR); since then, considerable progress has been made in creating a commonly accepted CBP framework applicable across all functions and closely tied to the *National Defense Strategy*.[1] Our methodology builds on the Strategic Planning Guidance used to shape DoD planning and programming decisions. It results in an Acquisition Investment Management model that recommends acquisition investments across a broad range of capabilities.

The AIM methodology is a top-down approach that uses forecasted trends in the national security landscape to determine future capability requirements. It adjusts Army acquisition investments into the future to provide the Army a portfolio of materiel capabilities that address the range of challenges it is likely to encounter or that will seriously threaten U.S. national security in future years.[2]

---

[1]   U.S. Department of Defense (2005).

[2]   From U.S. Department of Defense (2004a).

Overview of Methodology

In general terms, the methodology adapts an ABP process to create an Army acquisition investment portfolio that reacts to estimates of the future national security landscape. A flowchart of our methodology is shown in Figure 2.1. Input steps are shown in orange, analysis steps are shown in yellow, and results are indicated in green.

- The ABP process, shown in the top row, starts with a series of fundamental assumptions guiding Army acquisition planning and a list of signposts that, if they emerged, would threaten the validity of these assumptions. Following the yellow boxes down the right side of the figure, the APB process next requires a series of shaping and hedging actions—investments—that could be used to prevent the emergence of a signpost (shaping actions) or mitigate its effect on the fundamental assumption in the event it emerges (hedging actions). These investments are described both as type (procurement, research and development (R&D), or recapitalization/modernization) and as a capability category (different functions/activities required of the Army).
- In parallel with the more traditional ABP steps, the AIM methodology incorporates a threat and likelihood/level of Army involvement (T&L) landscape (indicated by the second and third orange boxes on the left of the figure and the second and fourth orange boxes in the next column to the right). This landscape is generated using various inputs: guidance from senior leaders, national intelligence estimates (NIEs), and the Strategic Planning Guidance. The T&L landscape includes a mixture of circumstances the Army must be prepared to face throughout the planning period. The AIM methodology also adapts the more traditional ABP by associating the ABP signposts with the circumstances that describe the anticipated national security environment (indicated by the bottom orange box on the left). The AIM methodology assumes that as a given circumstance becomes more likely and its threat to U.S. national security becomes more severe, the utility of investment actions that provide Army capability to address the circumstance is greater than the utility of investing

Figure 2.1
Flow Chart of Methodology

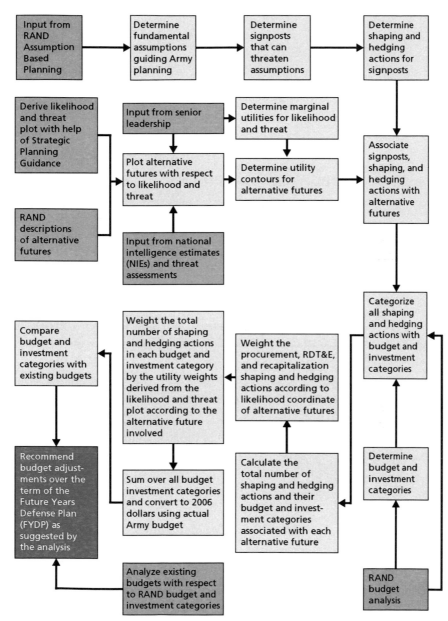

against circumstances that are not at this level of threat or likelihood (series of yellow boxes in the bottom half of the figure).[3]
- The results of the analysis are combined with an analysis of existing budget and investment categories (leftmost orange box on the bottom) to produce a portfolio of investments that provides a mix of materiel capability for the Army.

The steps used in this process will be described in greater detail throughout this chapter.

Applying the Traditional ABP Process

Classical ABP begins with three steps: (1) identifying fundamental assumptions, (2) developing signposts, and (3) developing shaping and hedging actions.

Identifying Fundamental Assumptions

Step one in ABP is the development of fundamental assumptions. We initially developed six assumptions for the Army's acquisition program:[4]

1. The U.S. Army will maintain operational and technological dominance over potential foes.
2. The U.S. Army will require and maintain a capacity for rapid global deployment and self-sustainment in austere theaters.
3. The U.S. Army will be increasingly adroit at managing complexity.
4. U.S. Army budgets will sustain operational and technical dominance.
5. The U.S. Army will rely on the capabilities of the Reserve Component and its sister services.

---

[3]   The utility function used in AIM to describe this relationship is based on the Cobb-Douglas class of utility functions. See Eaves (1985, pp. 226–239).

[4]   The development of fundamental assumptions was coordinated with personnel from the Office of the Assistant Secretary of the Army (Acquisition, Logistics, and Technology).

6.   The U.S. Army's enemies will span the range of capabilities but will all be competent and adaptive.

We used only the first five of these assumptions in the traditional ABP process. These five are directive and therefore form the core assumptions for developing signposts and, subsequently, hedging and shaping actions (investments). In contrast, the sixth assumption is qualitatively different. It is not directive and, therefore, acquisition hedging and shaping actions would not directly affect the assumption's validity. Additionally, the Army would not want to take hedging and shaping actions to protect the validity of the assumption, since it would be preferable if this assumption were invalid. Although we did not include this assumption in the traditional ABP process, we incorporated it into the AIM methodology through the development of the T&L charts in which the range of potential adversaries is built into the various circumstances plotted on the chart. The process for doing this is explained below.

Developing Signposts

For each of the five assumptions identified in the previous step, we next developed a number of signposts that would, if they were to appear, indicate potential vulnerability of the assumption. These assumptions and signposts are listed, in somewhat condensed form, in Table 2.1 and are detailed in Appendix A.

Developing Shaping and Hedging Actions

The final step in the traditional ABP portion of the AIM methodology is to develop shaping and hedging actions that either protect the validity of the fundamental assumption or help manage the consequences when the validity is challenged. For the AIM methodology, hedging and shaping actions take the form of acquisition investments, including investments that

- procure new materiel (procurement)
- invent and develop new materiel (R&D)
- refurbish and sometimes modernize existing materiel (recapitalization).

Table 2.1
Example: Assumptions and Associated Signposts of Vulnerabilities

1	**The U.S. Army will maintain operational and technological dominance over potential foes**
	a Potential peer adversaries make significant investments in high-end military capability
	b Potential regional adversaries demonstrate an intention to dominate a region militarily and a growing military capability
	c Nuclear weapons and delivery systems proliferate
	d The U.S. defense industrial base deteriorates
	e Countermeasures to important U.S. military technologies appear and begin proliferating
	f New weapons or doctrine based on new technologies appear outside the U.S. military
2	**The U.S. Army will require and maintain a capacity for rapid global deployment and self-sustainment in austere theaters**
	a Potential adversaries invest in or deploy significant anti-access capability
	b Host nation support becomes questionable or contingent on the nature and location of U.S. operations
	c U.S. foreign policy becomes isolationist
	d The U.S. military is refocused on homeland defense
	e Rapid deployment of U.S. Army forces is technically unavailable or unaffordable
	f National policy does not require rapid Army deployment
	g U.S. interest in austere theaters wanes
3	**The U.S. Army will be increasingly adroit at managing complexity**
	a The Army remains too operationally engaged to train and experiment effectively
	b Very sophisticated enemies emerge that can challenge the Army's information dominance
	c The Army cannot recruit the right kind of people
	d Numerous operational, planning, logistic, and intelligence failures by the Army occur over a short period of time
	e High-tech units fare poorly at the National Training Center/Joint Readiness Training Center and do not improve over time
	f Technology readiness levels (TRLs) of critical technologies mature too slowly
	g Systems integration proves increasingly difficult
4	**U.S. Army budgets will sustain operational and technical dominance**
	a Congressional support for large defense budgets wanes
	b Operations tempo (OPTEMPO) continues to consume a large percentage of the Army budget
	c Re-missioning allocates resources in favor of other services
	d Personnel costs continue to increase significantly
5	**The U.S. Army will rely on the capabilities of the Reserve Component and its sister services**
	a OPTEMPO challenges the ability of reserves to train
	b Multiple contingencies divert the attention of the Reserve Component and sister services from Army priorities
	c Sister services allocate their resources in ways that do not support Army priorities
	d Other services and the Reserve Component are unable to recruit and retain sufficient numbers of the right kind of people
	e Force structure reductions reduce the number of contingencies other services can manage

Additionally, investments must be organized in a way that characterizes how the Army will manage the circumstances it must be prepared for. Therefore, we created investment categories to help the Army manage its planning and programming decisions. We chose ten investment capability categories, described in Table 2.2.

Table 2.2
Army Investment Capability Categories

Investment Capability Category	Description
1. Close battle	This budget category includes accounts that support close battle: the fight inside 400 meters. It includes such equipment as individual weapons, body armor, night vision goggles, and other items that contribute to the close fight.
2. Mobility	This budget category includes accounts that produce mobility. If a piece of equipment's primary function is the tactical movement of men or materiel (e.g., armored personnel carrier, five-ton truck, UH-60 helicopter), it falls into this category. The fact that these platforms are typically armed for self-defense does not move them from this category into any other (direct fire, forward support).
3. Direct fire	This budget category includes accounts for direct fire weapons that are not included in the close battle category, including larger-caliber cannon systems and anti-tank guided missiles. The category also includes tanks and similar vehicles that serve as platforms for direct fire gun and missile systems.
4. Indirect fire	This budget category includes accounts for indirect fire weapons: mortars of all calibers, artillery and non-line-of-sight rocket/missile systems.
5. Forward support	This budget category includes diverse accounts meant to provide combat and combat service support forward in the theater of operations. It includes engineer systems, logistics support, medical support, and generally any capability found in forward support battalions, Division Support Commands (DISCOMs), and Corps Support Commands (COSCOMs).
6. Force protection	This budget category includes accounts meant to provide force protection, ranging from air defense systems to immunizations to barrier materials. If the primary function of a system contributes directly to force protection, it is included in this budget category.
7. Remote support	This budget category includes accounts that support Army forces from afar. It includes the Training and Doctrine Command (TRADOC) school system, distance learning, and training support. It also includes the support provided by depots in the continental United States (CONUS), laboratories, analytical centers, and all of the combat support and combat service support provided from somewhere outside the immediate theater of operations.

Table 2.2—Continued

Investment Capability Category	Description
8. Command and control	This budget category includes accounts that provide command and control capabilities. It includes communication systems, battle management systems, situational awareness systems, and data display systems that contribute to a commander's ability to understand the battlespace and maneuver forces over and through the battlespace for advantage over the enemy.
9. Reconnaissance, surveillance, and target acquisition (RSTA)	This budget category includes accounts that provide reconnaissance, surveillance, or target acquisition capabilities. It includes radars, ground sensors, aerial sensors, and intelligence production, fusion, unmanned aerial vehicles, and transmission systems, e.g., the All Source Analysis System (ASAS), and the Special Operations Command Research Analysis and Threat Evaluation System (SOCRATES).
10. System integration	This budget category includes accounts that contribute to systems integration.

An example of this categorization is shown in Table 2.3. In this table, we examine the first assumption involving the U.S. Army maintaining operational and technological dominance over potential foes, and signpost A(iii) (see Table 2.1), which involves an adversary investing significantly in naval and anti-naval capabilities. We then list the shaping and hedging actions (acquisition investments) needed to counter the advent of this signpost and show the categorization of these actions in terms of budget and investment types. For reference, the insets show the budget and investment categories. For this analysis, the emphasis of the hedging and shaping investments is characterized as either reduce, invest, increase investment, or significantly increase investment.

Identifying Alternative Circumstances Requiring Army Capabilities

Under more traditional ABP, the development and application of hedging and shaping actions completes the formal process. The AIM methodology, however, uses the hedging and shaping actions as tools for

Table 2.3
Example of the Categorization of the Shaping and Hedging Actions

Assumption	Signpost	Shaping Actions	Hedging Actions	Budget Category	Investment Category
1. The U.S. Army will maintain operational and technological dominance over potential foes	A. Potential peer adversaries significantly intensify their investments in one or more of the following iii. Naval and anti-naval capabilities that threaten the U.S. Navy command of the oceans				
		Invest in remote support equipment		7	1
		Invest in forward support equipment		5	1
		Invest in force protection		6	1
			Invest in close battle R&D	1	2
			Invest in mobility R&D	2	2
			Invest in direct fire R&D	3	2
			Invest in close battle R&D Future Combat System (FCS)	1	2
			Invest in forward support R&D (FCS)	5	2
			Recapitalize and modernize existing close battle equipment	1	3
			Recapitalize and modernize existing mobility equipment	2	3
			Recapitalize and modernize existing direct fire equipment	3	3
			Recapitalize and modernize existing indirect fire equipment	4	3
			Recapitalize and modernize existing forward support equipment	5	3

Investment category	Description
1	Procurement
2	RDT&E
3	Recapitalization

Budget category	Description
1	Close battle
2	Mobility
3	Direct fire
4	Indirect fire
5	Forward support
6	Force protection
7	Remote support
8	Command and control
9	RSTA
10	Systems integration

developing an acquisition portfolio, so the analysis must be able to prioritize investments by type (procurement, R&D, recapitalization) as well as by capability category. This means that we need to assess what the Army must be prepared for. This is accomplished by developing a set of alternative circumstances for which the national leadership is likely to depend on Army capabilities, if a response to the circumstance is required.

The alternative circumstances we consider in our analysis are listed in Table 2.4. More detailed descriptions and our assumptions about the Army role in dealing with each set appear in Appendix B.[5]

With an appropriate set of alternative circumstances established, the next task in the AIM methodology is to locate the set onto a map of the national security environment. Since this mapping must allow prioritization of the Army's response to the alternative circumstances, we chose a plot that included the threat that the circumstance represents to the national security on the vertical axis and the likelihood/level of Army involvement on the horizontal axis.

Our initial assessments of the locations of the alternative sets of circumstances, as well as their movement over time, in the T&L plot are shown in Figure 2.2.[6]

The locations of the alternative sets of circumstances in the T&L plot represent a top-level view of previous, present, and future world situations parsed along the lines of the alternative conditions we have described. We located these alternative circumstances in the figure using our best personal knowledge.[7] In a review by senior Army

---

[5] Other users of the AIM methodology may prefer alternative sets of circumstances. For example, disaster relief may become a more important Army responsibility. Since acquisition investment recommendations flow from the set of circumstances, however, it is important that whatever set is chosen be acceptable to the Army leadership. Leaders must agree that the set broadly defines the circumstances for which the Army must be prepared.

[6] In this exercise of the AIM methodology, 2004 is the baseline year. We attempted an exercise in 20/20 hindsight and placed the circumstances on the T&L plot as we believe we would have plotted them in 2003, when the 2004 acquisition budgets were being formulated.

[7] The plots also contain a set of circumstances labeled as 2008 Peace. For this set, we conjectured circumstances where the world becomes significantly less dangerous. For example, we envisioned that several countries currently intent on gaining a nuclear arsenal agree to

Table 2.4
Alternative Circumstances Considered in the Analysis

Alternative Circumstance	Description
Regional insurgency	Identity politics fuel local trouble. This future is characterized by ethnic warfare, expulsions, and scenes such as those seen in the Balkans.
Regional anarchy	Governmental control in an area has eroded to the extent that gangs or warlord armies dominate large regional areas. This creates an environment in which terrorist and criminal activities are easily concealed. It promotes large-scale humanitarian crises such as famines, pandemics, and refugee exodus to neighboring countries.
Emergent nuclear powers	These circumstances focus on newly nuclear powers (e.g., North Korea, Pakistan, perhaps a Brazil or Iran) with immature command and control, little or no strategic warning, poor intelligence, surveillance, reconnaissance (ISR), and situational awareness that might lead to accidental, unauthorized launches, or launches under misunderstood or mistaken circumstances.
Cold peace competition	These circumstances include ideological differences based on nationalism, religion or politics, or cleavages of civilization. These circumstances might find China, Russia, the United States, or perhaps a unified Europe, concluding that their interests are too divergent to warrant cooperation in international affairs and might motivate some of the states to interfere and subvert others as they pursue their global interests and objectives. Miscalculation or frustration by one party could lead the group to blunder into war, so the future is very dangerous.
Broad Islamic insurgency and terrorism	These circumstances involve two main elements: (1) campaigns to undermine secular governments within North Africa, the Middle East, and throughout the Muslim Crescent extending through South Asia into Southeast Asia and to replace them with Islamist, theocratic regimes, and (2) punitive attacks on the West and especially the United States.
Rising conventional power	A state such as Iran or Japan imperils an important U.S. goal or interest for the region (e.g., undermines U.S. stability operations in Iraq) or endangers U.S. access to crucial resources.

put aside those ambitions. This alternative set of circumstances was constructed to test AIM's sensitivity to large changes in the national security environment.

Figure 2.2
Threat and Likelihood Plot of the National Security Environment Between 2004 and 2008

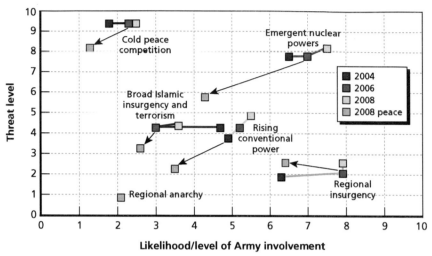

RAND *MG532-2.2*

leadership, the locations would be based on NIEs or other intelligence forecasts as well as on Army planning criteria for involvement in the contingencies presented by the alternative circumstances.

We plotted the regional insurgency alternative in the lower right quadrant of the figure for 2006 because such conditions would not pose much of a threat to the United States but could find the Army rather deeply involved. This illustrates the dual aspect of the horizontal axis in capturing both the likelihood of the alternative as well as the level of Army involvement. We felt that the level of threat to the United States from this alternative would increase slightly in 2008.

The regional anarchy alternative is plotted low on the likelihood and threat axes because these very local disruptions are most often handled by other nations more directly involved with the region and because this alternative in its current incarnation does not pose a substantive threat to the United States. We felt that the likelihood and level of threat of this alternative would remain the same in 2008.

The emergent nuclear powers alternative for 2006 was plotted rather high on the vertical threat axis because we felt that such a circumstance would be very dangerous to the United States. Likewise, we felt that the Army's involvement in such an alternative would be very high, and accordingly we placed this alternative high on the likelihood/involvement axis. Furthermore, we note that this alternative world is emerging even as this monograph is being written and that the threat and likelihood of this alternative would increase in 2008.

To locate the cold peace competition alternative in 2006 on the plot, we decided that the degree of Army involvement would vary with the specific circumstances but we concluded that the likelihood of this circumstance is quite low. For 2008, we felt that the threat level would remain the same although the likelihood increases slightly.

When determining where in the plot to locate the broad Islamic insurgency and terrorism alternative in 2006, we could not conceive of credible circumstances in which a state government would provide much active support for terrorist weapons of mass destruction (WMD) and concluded that even regimes such as that in North Korea would understand that giving WMD to such groups would pose dangers to itself. The threat coordinate reflects our judgment that attacks such as 9/11, Madrid, London, and Bali, although devastating, do not threaten the existence of the state (although the Madrid attack appears to have changed the election outcome for Spain). Unless something changes considerably, this particular set of conditions would not present a threat that can mount a serious campaign with a series of sequenced attacks that could threaten U.S. society and the Republic. The likelihood coordinate reflects the outcome of a thought experiment that we conducted: If we disconnect Iraq from the war on terrorism and treat it more as the regional troublemaker it is, Army involvement in the global war on terror looks much smaller: foreign internal development and capacity-building among beleaguered countries, modest deployments of troops in Afghanistan, special operations forces (SOF) in the Philippines and elsewhere, but no major ground force deployments. For 2008, we saw a slight increase in the likelihood and threat coordinates.

When determining where in the plot to locate the rising conventional power alternative in 2006, we felt that such a future could impose a significant level of threat and could require a much more robust response from the Army—almost certainly more forces than would be involved in the broad Islamic insurgency and terrorism alternative. For 2008, we felt that both the threat and the likelihood for this alternative would increase.

Prioritizing Alternative Circumstances for Budgeting

The T&L plot now provides a useful tool for prioritizing the circumstances for budgeting purposes. The prioritization is accomplished using a utility function. We have adopted the standard Cobb-Douglas utility function for our analysis:

$$U(x_1, x_2) = x_1^{1/2} x_2^{1/2}$$

where x_1 is the likelihood/level coordinate value in our plot and x_2 is the threat axis coordinate value. In this formulation the marginal utility of a change in threat is the inverse of marginal utility of a change in likelihood. This leads to the utility contours asymptotically approaching the axes.[8]

Figure 2.3 illustrates the contours of constant utility on our T&L plot along with the positions of the alternative circumstances. In our model, increasing budget priority is given to alternative circumstances that are rated more highly in this utility plot. According to the equation shown above, the utility increases from the lower left-hand corner of the plot to the upper right-hand corner, as indicated in the utility contour values along the top edge of the plot. For example, all the shaping and hedging actions in the budget categories associated with the emergent nuclear powers alternative would get a higher priority than those associated with the regional anarchy alternative. This is consistent with our earlier assertion that the Army should focus investments in the resources needed to address situations that are more likely to

---

[8] Gale (1960). Senior Army leaders may feel that threat changes are more significant than likelihood changes. Changing the marginal utilities of the two axes to reflect such a judgment is straightforward and standard in applications of utility functions.

Figure 2.3
Utility Contours

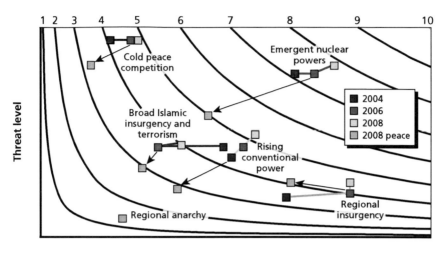

Likelihood/level of Army involvement

occur, that would involve the Army in a significant way, and that pose a significant threat to the United States.

We believed that the world would be a more dangerous place in 2008 than in 2006. This is reflected in the utility contours in Figure 2.3, and the effect on the budget priority for the alternative circumstances in our model is evaluated in Figure 2.4. As shown in the latter figure, the priority for the emergent nuclear powers alternative is the highest for both model years, whereas the priority for the regional anarchy alterative is the lowest for both model years. For all the alternatives, the budget priority increases in 2008 from its priority in 2006 with the exception of the regional anarchy alternative, which stayed the same.

Figure 2.4
Acquisition Budget Priorities of the Alternative Circumstances Based on Utility Contours

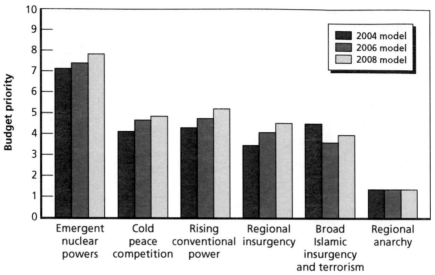

RAND MG532-2.4

Identifying Investments in Need of Emphasis

As shown in Figure 2.1, the ABP and the T&L paths in the methodology intersect with the task of associating the signposts and the shaping and hedging actions with the alternative circumstances. After comparing the descriptions of the signposts with the descriptions of the alternative worlds, we concluded that certain signposts are most likely to emerge given specific changes in the alternative worlds. Table 2.5 maps the associations between alternative circumstances (Table 2.4), load-bearing assumptions about Army acquisition (Table 2.1), and the signposts of vulnerability (Table 2.1). For example, under assumption 3, *the Army will be increasingly adroit at managing complexity*, signpost d (*"The advent of numerous operational, planning, logistic, and intelligence failures by the Army over a short period of time"*) can be considered significant under all the alternative futures. However, for this same

Table 2.5
Associations Between Alternative Circumstances, Fundamental Assumptions, and Signposts of Vulnerability

Alternative Future	Load-Bearing Assumption				
	1. U.S. Army will maintain technical and operational dominance over competent and adaptive foes	2. Require and maintain a capacity for rapid global deployment and self-sustainment in austere theaters	3. Increasingly adroit at managing complexity	4. Budgets will sustain operational and technical dominance	5. Rely on capabilities of the Reserve Component and sister services
Regional insurgency		2b,2c,2d,2f,2g	3d	4a,4b,4c,4d	5a,5b,5c,5d,5e
Regional anarchy		2b,2c,2f,2g	3d	4a,4b,4c,4d	5a,5b,5c,5d,5e
Emergent nuclear powers	1b,1c,1e	2a,2b,2c,2e,2f,2g	3a,3c,3d,3e,3f,3g	4a,4b,4c,4d	5a,5b,5c,5d,5e
Cold peace competition	1a,1d,1e,1f	2a,2b,2c,2e,2f,2g	3a,3b,3c,3d,3e,3f,3g	4a,4b,4c,4d	5a,5b,5c,5d,5e
Broad Islamic insurgency and terrorism	1c	2b,2c,2d,2f,2g	3a,3c,3d	4a,4b,4c,4d	5a,5b,5c,5d,5e
Rising conventional power	1b,1d,1e,1f	2a,2b,2c,2e,2f,2g	3a,3c,3d,3e,3f,3g	4a,4b,4c,4d	5a,5b,5c,5d,5e

NOTE: Load-bearing assumptions and signposts of vulnerability (in the body of the table) are labeled to match their placement in Table 2.1.

assumption, signpost b (*"The emergence of very sophisticated enemies who are able to purposely inject added complexity while simultaneously reducing the U.S. Army's ability to deal with complexity"*) is associated only with the cold peace competition alternative. The associations displayed in Table 2.5 were derived from many discussions among the research team.

Each signpost in Table 2.5 has a number of hedging and shaping actions associated with it. As explained above, the hedging and shaping actions are investments, specified by budget type and capability category. The level of investment associated with each hedging and shaping action provides a way to quantify the investment for AIM, as follows:

- –1 point: reduce
- +1 point: invest
- +2 points: increase investment
- +3 points: significantly increase investment.

The next step in the AIM methodology process is to sum the total number of investment points across each alternative circumstance in each of the investment and budget categories.[9] For example, across regional insurgency there were 17 close battle R&D points. Summing investment points in this manner provides a quantitative measure of overall investment emphasis by budget type and category for each alternative.

The next step in the AIM methodology adjusts the investment emphasis on R&D, procurement, and recapitalization for each alternative circumstance according to the circumstance's placement on the T&L plot. We assumed this dependence because in ABP, shaping actions are taken to prevent the emergence of signposts that threaten the fundamental assumptions. Hedging actions are taken to manage the consequences of assumption invalidity. This formalism suggests that RDT&E investments should be large when the likelihood of an alternative circumstance is low, whereas procurement and recapitaliza-

---

[9] We sum total investment points, rather than total hedging and shaping actions. As noted above, there are four levels of investment with –1 to 3 investment points.

tion investments should be large when the likelihood of the circumstance is high. To account for this feature of the methodology, the investment emphasis associated with a given alternative circumstance is weighted linearly according to the coordinate of that future along the likelihood/involvement axis—RDT&E decreasing with increasing likelihood; procurement and recapitalization increasing with an increasing coordinate.

A second weighting now applies the utility function value of each circumstance's position to the investment emphasis for each of its budget types and categories. This weighting adjusts the ratios of acquisition investments according to the anticipated national security environment.

To illustrate what the model results look like at this point in the analysis, Figures 2.5 and 2.6, respectively, show the results for the cold peace competition and the emergent nuclear powers alternatives in 2006. Comparing these figures, we see that the cold peace circumstance has a significantly lower overall priority than the emergent nuclear powers circumstance. This difference is the result of each circumstance's respective placement on the T&L plot and its very different utility function values. Additionally, the budget emphasis across each category for the two alternatives is quite different, which is consistent with the varying shaping and hedging actions made for these circumstances. The cold peace competition circumstance gives greater relative emphasis to RDT&E, whereas the emergent nuclear powers circumstance emphasizes procurement and recapitalization, reflecting that circumstance's greater likelihood.

Pushing through to the final steps in the analysis, the investment emphasis for each budget category is summed across all alternative circumstances. This step aggregates the model outputs for the first time.

Figure 2.5
Total Number of Shaping and Hedging Actions in Each Budget and
Investment Category (Cold Peace Circumstance, 2006)

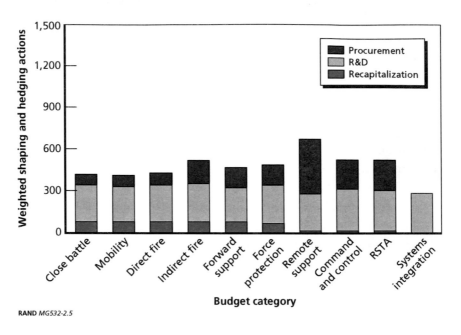

Normalizing the Model's Outputs

The final calculation in the AIM process normalizes the model's outputs by baselining them against an actual acquisition budget. We chose 2004 as our baseline.[10] Baselining from a year too far in the past could cause the overall acquisition budget recommendations emerging from the model to be too different from current budgets to be useful. This

---

[10] Choosing a baseline requires two assumptions: First, that the overall level of acquisition spending (procurement, R&D, recapitalization) is reasonable given the national security environment at the time; second, that the ratio of R&D spending to procurement and recapitalization is also reasonable. If, with hindsight, either of these assumptions proves inadequate, the analyst may develop and use any amount of acquisition spending and ratio of R&D to procurement/recapitalization as a baseline for the model.

Figure 2.6
Total Number of Shaping and Hedging Actions in Each Budget and Investment Category (Emergent Nuclear Powers Circumstance, 2006)

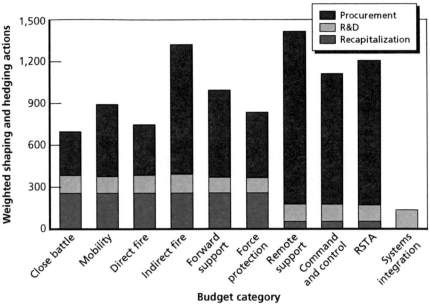

RAND MG532-2.6

is particularly the case when events, unanticipated at the time of budget formulation, significantly alter the national security environment for which the Army is preparing.[11] Additionally, 2004 was near enough in time so that the collective memory of the study group had a reasonable chance of reconstructing the national security landscape in a way that would mirror the thinking of the time.

The first normalization adjusts the ratio of the R&D investments to the procurement and recapitalization investments.[12] This simple cal-

---

[11] For example, using a FY 2000 budget as the baseline would cause the model to produce lower budgets than are the norm today, because the Army's overall budget was significantly increased following the September 11, 2001, terrorist attacks.

[12] This ratio is important because R&D is qualitatively different from procurement and recapitalization and the costs associated with these activities are different. R&D develops new materiel, whereas the other budget types produce and field materiel.

culation merely reduces or enlarges the R&D investment proportion so that the ratio of the R&D to the procurement/recapitalization matches that ratio in the baseline year. At this point in the analysis, the model has produced the final investment emphasis in each budget type and category, allowing the analyst to examine the overall structure of the acquisition budget for the first time.

The second normalization turns the final investment emphases into budget dollars by weighting the sum of the investment emphasis for each budget type and category so that the total equals some total dollar figure acceptable to the analyst. For the current analysis, we weighted the FY 2004 model outputs so that the overall model budget equals the actual FY 2004 acquisition budget.

At the aggregate level, the comparisons between the actual and modeled budgets are illustrated in Figure 2.7.[13] The important comparison here shows that the model recommends a slightly larger Army acquisition budget than is currently planned ($28 billion vs. $26 billion in FY 2008). This rise reflects the fact that we viewed the world as becoming somewhat more dangerous in 2008 than it was in 2006. Overall, however, the differences between the model and the actual budgets, at the aggregate level, appear reasonable.

A comparison of our investment model and the actual Army budgets for 2006 and 2008 is shown in Figures 2.8 and 2.9, respectively.

In these figures, the budget category is plotted horizontally and the budget amount in millions of 2006 dollars is plotted vertically. The first stacked bar in each figure (solid colors) is the result of AIM and the second stacked bar (striped colors) is the actual Army budget. Red bars represent procurement, green bars are RDT&E, and the blue bars are the recapitalization category.

To gain insights from Figure 2.9, we look for areas of significant difference between modeled budget recommendations and the actual 2006 Army acquisition budget. In two areas, mobility and systems integration, the actual budget is substantially larger than that

---

[13] Actual budgets do not include the supplemental budgets used to fund operations in Iraq and Afghanistan.

Figure 2.7
Total Army Acquisition Budgets (Actual and Modeled)

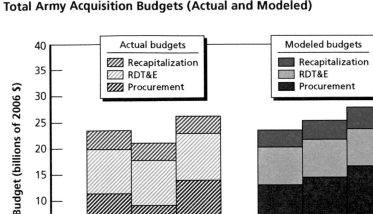

recommended by the model.[14] The large difference in the mobility category reflects, to some degree, ongoing Army operations. As trucks, Humvees, and helicopters are used heavily and the wear-and-tear takes its toll, they need to be rebuilt or replaced. The large difference in "systems integration" most likely represents the Army's commitment to the Future Combat System (FCS). Both of these examples demonstrate that this model is not meant to be deterministic or to provide "The Answer." Rather, the model acts as a guide to future acquisition investment based on anticipated future security environments. Other factors, such as ongoing operations and transformation, will also need to be considered as the investment strategy is developed.

Other budget categories, including close battle, indirect fire, and RSTA, are emphasized more in the model than in the actual budget. What this suggests is that for the national security

---

[14] In part, this may be definitional. The level of detail in the budget documents used to display the actual budget provided only a high-level view of some large budget items we marked as "mobility" and "systems integration." More detail might allow these to be better allocated among the budget categories.

Figure 2.8
Comparison of Actual and Modeled 2006 Army Acquisition Budgets

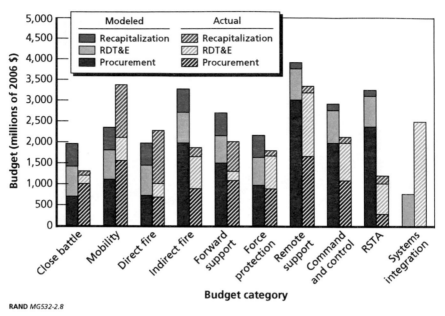

landscape we posited, these budget categories are more important than the current acquisition budget indicates.

When we consider the more detailed investment categories of procurement, RDT&E, and recapitalization, we also find some differences between the actual budget and our investment model, but these seem reasonable for the most part. For example, the proportion of R&D in the close battle budget category is significantly larger in the model than in the actual budget. This difference is most likely a function of the emphasis we placed on improving close battle capabilities to deal with the challenges of the posited national security environment. The actual recapitalization budget for mobility is larger in the actual budget and, as mentioned above, this likely reflects the needs of ongoing operations.

Referring to the comparison of our investment strategy and the actual 2008 budget shown in Figure 2.9, we see that there is again a general agreement between the actual 2008 budget and our recommended budget, although the modeled budget is slightly larger overall.

Figure 2.9
Comparison of Actual and Modeled 2008 Army Acquisition Budgets

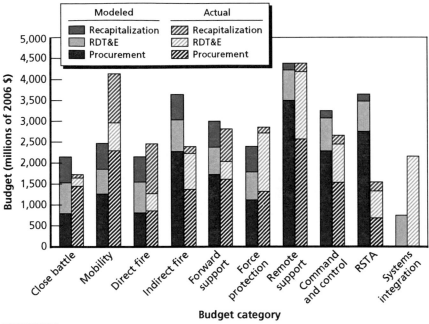

CHAPTER THREE

Integrating AIM into Army Planning and Programming for Materiel Acquisition

Acquisition programs must operate within the parameters established by the Defense Resources Board and the Secretary of Defense through the PPBS.

— *U.S. Army War College (2005, p. 200)*

The AIM methodology offers a way to monitor and adjust the Army's materiel-acquisition plans so that they are better able to keep up with changes that may be occurring in the national security environment. AIM is designed to work with the Army's current acquisition processes and organizations. In this chapter, we discuss how AIM fits (or could fit) into the official planning, programming, budgeting, and execution (PPBE) process the Army uses to allocate all its resources—acquisition and otherwise.[1]

We focus on planning and programming—not on budgeting and execution—because AIM is designed to help increase the strategic responsiveness of materiel acquisition in the Army, not its tactical responsiveness. The former is the province of planning and programming; the latter the province of budgeting and execution. To be sure, if major changes occur very suddenly in the external environment, and

---

[1] The Army PPBE process is the Army's portion of the DoD-wide planning, programming, budgeting, and execution system (PPBES). The DoD PPBES is a formal resource-allocation, tracking, and control system that DoD has employed (in one form or another) for more than 40 years, ever since the system was first established within the department by Secretary of Defense Robert McNamara in the mid-1960s.

no signposts have triggered, indicating that they were coming, budget and execution changes may be necessary, and AIM could be used to evaluate whether these changes make sense. But under normal conditions, planners and programmers should be able to use AIM to monitor signposts, thus diminishing the need for any late-breaking, "catch-up" adjustments.

The chapter proceeds by reviewing how the Army currently does planning and programming for materiel acquisition, including which offices and organizations are involved and how they interact with one another. In the course of that review, the chapter describes where and how AIM can fit naturally into the process.

We emphasize, however, that AIM is not meant to be a determinative tool in the planning and programming process. As noted at the end of this chapter, there may be a number of reasons to allocate acquisition resources differently than suggested by AIM and these other factors need to be taken into account. Also, since predicting the future of the global security environment is necessarily an imprecise art, such predictions should be used for general guidance only. AIM's outputs should, therefore, be considered only as proportional estimates of resource allocation for the type of Army capabilities needed to address the predicted future global security environment.

An important underlying theme of the chapter is that if AIM is incorporated into Army planning and programming for materiel acquisition, senior Army decisionmakers will have access to a "portfolio management" perspective that the materiel-acquisition system has never made available to them in the past.[2] Traditionally, materiel acquisition in the Army has taken a program-specific perspective, i.e., one that focuses on how particular materiel solutions address particular threats.[3] If AIM information is included as an overlay and supplement

---

[2] For more on portfolio management, see Davis, Kulick, and Egner (2005).

[3] Or, as they are now being called, particular "capability gaps." DoD has shifted from what used to be called "requirements development" to what is now called "capabilities development." Both terms refer to what happens at the "front end" of the materiel-acquisition process, when a need for new materiel systems is recognized and decided on, before the formal materiel-acquisition process itself—i.e., design, development, engineering, and production—begins.

to that program-specific information, decisionmakers will have important new information to use when setting priorities and allocating resources among all the individual systems and programs about which they have to make decisions.[4] AIM can provide them with information for assessing the general "correctness" of their resource planning decisions while suggesting new resource priorities in response to a changing national security environment.

Planning and Programming in the Army for Materiel Acquisition[5]

Many Army offices and organizations—all of which interact with the office of the Assistant Secretary of the Army (Acquisition, Logistics, and Technology (ASA(ALT))—are involved in the planning and programming for materiel acquisition in the Army. Because these offices and organizations must collaborate to produce the Army's materiel-acquisition resourcing plan, all are potential users of AIM output. In this section, we review the offices and organizations that are involved, what they do, and the nature of their interaction with the ASA(ALT)'s office. It is in the context of the interactions and overlapping responsibilities within and among these offices and organizations where natu-

---

[4] An additional benefit of incorporating AIM into Army planning and programming for materiel acquisition is that the Army will be conforming to the new DoD policy calling for a more strategic, "top-down" approach to materiel acquisition. That new policy is reflected in the Joint Capabilities Integration Development System (JCIDS) issued by the U.S. Joint Staff (2001) and the "Aldridge Study" on PPBS reform issued by the Office of the Secretary of Defense (OSD) (January 2005). (The latter study is cited in the current Army Strategic Planning Guidance (2003–2023) as the direction the Army will follow. It is in the Army's interest to use AIM in any case—Army leaders face the same kinds of resource-allocation challenges within the Army as DoD leaders face at their level—but if adopting AIM means that the Army is also following DoD's new policies for materiel acquisition, so much the better.

[5] For an extended discussion of DoD-level PPBE processes as they relate to materiel acquisition, see the *Defense Acquisition Guidebook* (DAG) (U.S. Department of Defense, 6th ed., 2006). The DAG accompanies DoD Directive 5000.1 and DoD Instruction 5000.2, which define DoD acquisition policy.

ral openings exist for AIM to be used as an additional tool to guide decisionmaking.

The RDAP, MDEPs, and G-8[6]

Organizational Responsibilities and Activities. The Army's Research, Development, and Acquisition Plan (RDAP) is a forward-looking, 15-year plan that the Army maintains and updates regularly as part of its PPBE process. At any given time, the RDAP provides a detailed view of what the Army intends to spend on the development and production of technologies and materiel over the next 15 years. The RDAP database contains a 1 – N priority list of RDT&E and procurement program packages ("resource buckets") called Management Decision Packages (MDEPs) with funding streams for an entire 15-year planning period. The 15-year planning period consists of the six years in the Future Years Defense Plan (FYDP), followed by the Extended Planning Period (EPP), which covers the nine years beyond the FYDP.[7] As time moves forward in the biennial planning and programming process, the two years at the "front" of the FYDP peel off to become the budget years (i.e., the two years for which a budget must be prepared), and the two years at the "front" of the EPP become part of the new FYDP.

The RDAP database for the FYDP and EPP is prepared and maintained by the Deputy Chief of Staff, G-8, along with the ASA(ALT), who also co-chair the "Equipping" (EE) Program Evaluation Group (PEG), a corporate Army body that sets the proposed dollar amounts in the materiel-acquisition MDEPs in the RDAP. Those amounts are subject to further review and decisionmaking by higher-level corporate forums in the Army as well. (The EE PEG and higher-level corporate forums are discussed further below.)

---

[6] The information in this subsection is drawn from U.S. Army War College (2005a).

[7] Under DoD's biennial approach to planning and programming, the FYDP alternates each year from covering six years (in even numbered years) to five years (in odd numbered years). As a result, the RDAP alternates each year from covering 15 years to 14 years, dropping two years off the front and adding two years at the back every other year, For example, the FY 2006 RDAP covers FY 2006–FY 2020, whereas the FY 2007 RDAP will cover FY 2007–FY 2020; the FY 2008 RDAP will then cover FY 2008–2022, and the FY 2009 RDAP will cover FY 2009–2022.

Taken together, MDEPs account for all Army resources, so they cover capabilities programmed for the Active Army, the Army National Guard, the Army Reserves, and the civilian workforce in the Army. Looking across all Army resources (not just those for materiel acquisition), an individual MDEP records the resources needed to achieve an intended outcome for a particular organization, program, or function. Each individual MDEP applies uniquely to one of the following:

- missions of modified tables of organization (MTOE) units
- missions of tables of distribution and allowances (TDA) units
- acquisition, fielding, and sustainment of weapon and information systems (with linkages to organizations).

The MDEPs relevant to the AIM method fall within this category of MDEPs:

- special visibility programs
- short-term projects.

Where AIM Might Be Used. Because of its "rolling" 15-year time horizon, the nature of the "resource buckets" it uses, and the fact that it is jointly maintained by the Deputy Chief of Staff G-8 and the ASA(ALT), the RDAP serves as the natural vehicle for integrating AIM into Army planning and programming for materiel acquisition.[8]

The office of the ASA(ALT), to help support G-8 in the latter's responsibility to maintain and update the RDAP database, could perform AIM "runs" to suggest where and how the dollar contents of the acquisition-related MDEPs might be changed in response to events in the projected security environment since the last Program Objective Memorandum (POM) cycle. The office of the ASA(ALT) would perform these AIM runs in time to support the regularly scheduled RDAP updates that G-8 makes each year, whether

---

[8] The Army Deputy Chief of Staff G-8 is responsible for Programming, Materiel Integration, and Program Analysis and Evaluation.

before a "full" POM-build in an even year or a "POM-update" in an odd year.

A critical characteristic of MDEPs that must be present if AIM is to work as intended has to do with the Army's authority (specifically Army Headquarters) to change MDEP values in the FYDP and EPP years. In particular, when assigning dollar amounts to MDEPs in the FYDP out-years (the FYDP years beyond the first FYDP year, which is the "budget year"), Army Headquarters is restricted only by Total Obligation Authority (TOA) limits, not by individual appropriation limits. This is important because AIM assumes complete "reallocation flexibility" across the investment categories of RDT&E, procurement, and recapitalization. Those categories relate directly to the appropriation categories of RDT&E, procurement, and operations and maintenance (O&M). For the budget year (but not for the FYDP out-years or the EPP years), funding allocations in those categories are subject to appropriation-specific constraints imposed by Congress to ensure that appropriated dollars are spent in accordance with Congress's wishes.

Further conducive to adopting AIM, MDEPs are grouped into Budget Operating System (BOS) categories that represent a common battlefield function or a common activity of the supporting Army infrastructure (e.g., aviation, ammunition). Most BOS groups are managed by a G-8 division. The division chief (known as the BOS manager), assisted by his or her staff and an ASA(ALT) counterpart, is responsible for overseeing the contents of the MDEPs within the BOS. That is, BOS managers are responsible for monitoring, facilitating, and recording the results of the deliberations and tradeoffs that take place within the "corporate" forums the Army uses to allocate resources within its internal planning and programming process. For materiel acquisition—i.e., RDT&E, procurement, and recapitalization—that allocation begins with the EE PEG.

The ten "bins" in the AIM methodology have been defined to be similar to the BOS groups the Army already uses but with some increased detail to support the AIM approach. Our assumption is that the BOS structure currently in place in G-8 could be adapted to match whatever set of AIM "bins" the Army ultimately decides to use.

G-3[9]

Organizational Responsibilities and Activities. The Deputy Chief of Staff (DCS) G-3 "prepares the Army's Strategic Planning Guidance (ASPG)," "defines Army planning assumptions," and "sets requirements and priorities based on guidance from the SecDef, the Secretary of the Army, the Army Chief of Staff, and the combatant commanders."[10] The DCS G-3 is also responsible for managing the planning phase of the Army PPBE process. In particular, G-3 "guides the work of the PEGs on planning matters, to include requirements determination (i.e., capabilities development) and prioritization." On the programming side, the DCS G-3 "assesses capabilities, deficiencies, and risks of the Program Objective Memorandum (POM) force at the end of the current POM."

Where AIM Might Be Used. Given its multiple roles, G-3 might benefit from using AIM in various ways.

First, G-3 could benefit by having AIM information when developing the Army's Strategic Planning Guidance. Indeed, given its assignments, G-3 may want to play a central role in defining the "threat-likelihood framework," the Army operating assumptions, and the signpost indicators that make up the AIM methodology.

Second, the DCS G-3's role in the PPBE process also is conducive to the use of AIM output. If MDEP values are changing in the EPP years, G-3 can use AIM to evaluate whether the changes are consistent with the Army's basic operating objectives in the face of national security changes. G-3 could also use AIM to propose and initiate revised EPP allocations, if signposts are being triggered indicating that one or more of the Army's basic operating assumptions may be facing risk.

The AIM methodology could be used to help G-3 assess the capabilities, deficiencies, and risks of the POM force. Use of AIM would allow G-3 to test whether out-year resource allocations in the

---

[9] The information in this subsection is drawn from U.S. Army War College (2005a, pp. 203–205).

[10] The Army Deputy Chief of Staff G-3 is responsible for operations and planning.

POM for materiel acquisition are consistent with projected national security developments and the associated shaping and hedging investments that could be made to respond to them.

Army Corporate Forums[11]

Responsibilities and Activities. The Army uses what it calls "PPBES deliberative forums" to provide venues where all concerned parties are given a chance to argue for resources.[12] The Army's six PEGs— for manning, training, organizing, equipping, sustaining, and installations—serve as the initial set of corporate forums for this purpose. As noted above, the EE PEG is the relevant PEG for resource allocations relating to materiel acquisition.

Each PEG "sets the scope, quantity, priority, and qualitative nature of resource requirements, . . . monitors resource transactions, and . . . makes both administrative and substantive changes to assigned MDEPs. MDEP proponents, subject matter experts, and, as appropriate, representatives of commands and agencies participate in PEG deliberations." Permanent members of every PEG, include representatives from the Assistant Secretary of the Army, Financial Management and Comptroller (ASA(FM&C)), appropriation sponsors, program prioritizers, and requirements staff officers from G-3 and G-8 Program Analysis and Evaluation (PAE) program integrators. All PEGs are charged with helping HQDA functional proponents to

- "maintain program consistency, first during planning and later when preparing, analyzing, and defending the integrated program-budget," and to
- "keep abreast of policy changes during each phase of the PPBE process."

---

[11] The information in this subsection is drawn from U.S. Army War College (2005a).

[12] All three of the military departments in DoD use similar such forums. The Air Force, for example, refers to the forums it uses to do resource allocation as the Air Force "Corporate Structure."

The Army PPBE deliberative forums above the PEGs include the Council of Colonels, the Planning Program Budget Committee (PPBC), the Senior Review Group (SRG), and the Army Resources Board (ARB). The Council of Colonels is chaired by the Chief of the Resource Analysis and Integration Office in G-3, the Chief of the Program Development Division on PAE in G-8, and the Deputy Director of Financial Management and Comptroller in the Office of the Assistant Secretary of the Army (OASA) (FM&C). The Council of Colonels "packages proposals," "frames issues," and "otherwise coordinates matters that come before the PPBC." In its dealings with the EE PEG, the Council of Colonels would be a natural consumer of AIM products.

The SRG is co-chaired by the Under Secretary of the Army and the Vice Chief of Staff, Army (VCSA). The SRG is the senior-level forum to resolve resource-allocation issues but generally does not revisit decisions made at lower levels. Among its responsibilities, the SRG makes recommendations to the ARB regarding resourcing alternatives.

The ARB is chaired by the Secretary of the Army with the Chief of Staff of the Army as the vice chair. The ARB is the senior Army leadership forum in the PPBE deliberative forum hierarchy. It approves prioritization of all Army programs, including materiel-acquisition programs, and selects resource-allocation alternatives.

The PPBC has three co-chairs, one of whom presides over the forum depending on its subject matter—the Assistant G-3 for planning, the Director, Program Analysis and Evaluation (DPAE) for programming, and the Director of the Army Budget (DAB) for budgeting and execution. Among its responsibilities, the PPBC is charged with maintaining overall discipline of the PPBE process and making sure that Army policy remains internally consistent and that program adjustments remain consistent with Army policy and priorities.

Where AIM Might Be Used. The Army's deliberative forums for assembling the POM could make natural use of AIM in performing their development, oversight, and review roles in the POM-building process.

The EE PEG, which as noted above is co-chaired by the ASA(ALT) and G-8, would be a likely user of AIM.

The portfolio-management perspective AIM provides is well-suited to supporting both of the above objectives, so the PPBC is another natural consumer of AIM products. In particular, the PPBC could use AIM to test whether proposed changes in resource allocations for materiel acquisition are directionally consistent with new national security circumstances that may be emerging—and, if they are not consistent, to ask questions to determine why they are not and to possibly direct changes so that they will be consistent. Among the PPBC's leadership, the Assistant G-3 and the DPAE would also be natural users of AIM output.

At their levels, the SRG and ARB are most likely to be interested in AIM outputs that can be used to determine whether changes being made in Army resource allocations for materiel acquisition are directionally consistent with emerging circumstances. In particular, because so many factors can affect future resource-allocation plans, the most valuable application of AIM may be at the PPBC, SRG, and ARB levels, to give senior Army leaders a way to independently check to see if the Army's materiel-acquisition resourcing plans are consistent with new security challenges that may be emerging and the Army's basic operating policies about how it will deal with such challenges.

Scheduling AIM Analyses

The PPBE calendar that the Army follows each year, which aligns with DoD's overall PPBE calendar, determines when AIM runs would be done to check or adjust MDEP allocations in the RDAP. Revisions being made to the RDAP typically occur during preparation of the combined POM/Budget Estimate Submission (BES) (April to August) and the President's budget (October to January). During these periods, Army Headquarters adjusts the allocations in the FYDP years (the first six years of the 15 years covered in the RDAP), and the Army research, development, and acquisition community adjusts the final nine years in the EPP.

Given that schedule, the times of the year that would make the most sense for AIM analyses to be done would be midsummer and the December–January time frame. In midsummer, the FYDP and EPP allocations are being finalized for the POM/BES submission. And in December–January, the RDAP associated with the President's budget could be used as the basis for AIM analyses that the Army may wish to provide to the Office of the Secretary of Defense (OSD) as input for the Strategic Planning Guidance to be issued by the Secretary of Defense in that period to guide the next POM/BES development cycle.

For this schedule to work, it would be necessary to coordinate with cycles and schedules governing the production of national intelligence estimates.

Figure 3.1 illustrates how AIM might be used in the Army planning and programming process. The figure uses FY 2007 and FY 2008 as an example of the biennial schedule that defines the recurring PPBES cycle followed by the Army and the rest of DoD. Under the biennial approach, "full" POMs—i.e., POMs that address how every dollar is allocated—are prepared only in even calendar years. In odd calendar years, the full POM from the preceding year may be updated as necessary (so AIM has a natural role to play here, too, as the figure indicates), but the presumption is that many if not most of the planned resource allocations over the remaining five years of the six-year-long FYDP defined in the previous year's "full" POM cycle will stay the same.

The figure illustrates how the office of the ASA(ALT) could use AIM to help G-8 maintain and update the RDAP database for the POM. As the figure indicates, OASA(ALT) would perform these AIM runs in time to support the regularly scheduled RDAP updates that G-8 makes each year, whether before a "full" POM-build in an even year or a "POM-update" in an odd year. The figure also illustrates G-3's role in providing strategic planning guidance to the Army planning and programming process, and how G-3 could benefit using AIM outputs to help support the Army's input to the DoD-level Strategic Planning Guidance, which the Secretary of Defense issues to drive overall DoD planning and programming. The figure shows the use that the Army's "deliberative forums" for assembling the POM could make in performing their development, oversight, and review roles in the POM-build process.

Figure 3.1
Using AIM in Army Planning and Programming: FY 2010–FY 2015 Example

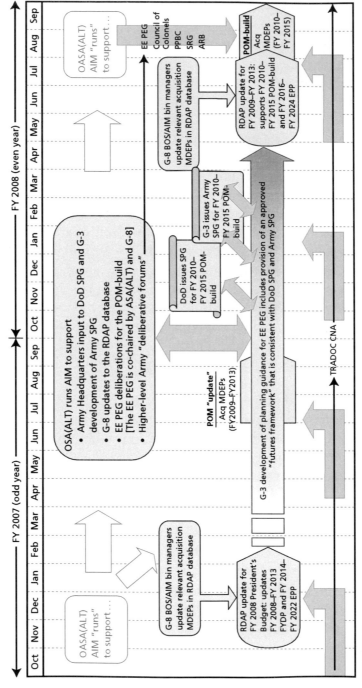

RAND MG532-3.1

Figure 3.1 also shows the capabilities needs analysis (CNA) that Training and Doctrine Command (TRADOC) performs to assess battlefield materiel capabilities and determine modernization alternatives for input to the Army POM. As described in HTAR (2005), the CNA is an interactive process among TRADOC's schools, proponents, and Army Headquarters staff. AIM outputs would have a natural role to play in informing those interactions and the conclusions they produce regarding how materiel-development resources should perhaps be reallocated if important changes are occurring in the projected national security environment.

Observations and Recommendations

We are aware that the proposed process relies on judgments about threats and likelihood and on acquisition officials to interpret intelligence reports and make decisions—decisions that some officials may believe are beyond their authority or that are at best the product of group judgment and consensus. The process may also seem to require the inputs of subject matter experts, many of whom may not reside in the acquisition community. We believe that such concerns can be properly addressed through carefully coordinated staff actions. The subsections below offer recommendations for implementing the acquisition investment strategy development process.

Form an Acquisition Strategy Working Group

The core of the acquisition investment strategy development process involves assessing the threat and likelihood of various circumstances and reprogramming funds into acquisition accounts in response. It would be helpful, therefore, to create an informal (although official) working group that would meet periodically to consider the location of these alternative sets of circumstances within the threat-likelihood plot and conduct the necessary assessments and recommend reallocations of funds across accounts in response to movement of some of those conditions. The working group might include members of the G-2 staff and other intelligence officers who can bring perspectives from the various elements of the intelligence community—warning, foreign technical intelligence, and other specialties—to bear on the question of new circumstances. The group might also involve representatives from the combatant commands, congressional liaison, and the Joint Capabilities Integration Development System (JCIDS) community so that the investment strategy development process could reflect expected advances in joint warfare and the effect of theater considerations: for example, the migration of key combat tasks into the Army or toward other services, and constraints imposed on warfighting by conditions prevailing in a given theater of operations.

Of course, there may be other legitimate reasons for maintaining high levels of investment in acquisition accounts that are not supported by our acquisition investment strategy development process: the health of the industrial base, for example. Therefore, it would be prudent to include representatives from the Program Executive Office (PEO) and perhaps a representative from OSD Industrial Policy in the working group so that these other perspectives in shaping acquisition investments receive due consideration.

Consider Other Alternative Circumstances

The different sets of circumstances posited in this monograph represent our translation of global security trends onto the Strategic Planning Guidance's threat-likelihood space. Other analysts might have better information and might populate that space somewhat differently. Moreover, as time passes, other concerns may become plausible as some factors in the international environment become more influential (e.g., the effect of energy competition on Indian-Chinese relations, the advent of new weapons types) or new factors appear—perhaps the emergence of a new regime with a revisionist foreign policy agenda, the unveiling of a new nuclear arsenal, or other events. Therefore, the Army acquisition community should periodically convene the acquisition strategy working group to consider new influential factors that could recast the threat-likelihood space. When meeting for this purpose, the working group might invite the relevant national and defense intelligence officers responsible for the regions and topics of concern to participate.

Exploit Assumption-Based Planning

As key Army planning documents evolve over time and as the planning considerations that underpin them change, it is possible that they will also involve new assumptions about the world and the Army's role in dealing with it. Therefore, it is important to review key plans periodically and to search them for indications of new or different assumptions. Where new assumptions are found, the acquisition community will want to generate new signposts and associated shaping and hedging strategies. The signposts and shaping and hedging should then be associated with alternative sets of circumstances to keep the acquisition investment strategy development process current.

Plan Acquisition Investment Strategy Reviews

The suggestions in the paragraphs above will take time to implement. Indeed, performing a diagnostic on Army plans in search of new, load-bearing assumptions can be time-consuming in its own right. Therefore, we recommend that the acquisition community establish a schedule, perhaps beginning in the off-budget year, to begin these activities. Because the international environment may be too volatile for Army planning to keep up with, there is a greater potential for different sets of international circumstances than for key, load-bearing assumptions. Thus, one possible approach would be to look for emergent alternative futures and to consider where alternative futures might lie in the T&L plot every other year.

CHAPTER FOUR

From the Lessons of the Past to the Potential of the Future

In conclusion, we note that in an earlier era, the Army struggled with many of the same issues that confront its acquisition strategy today. In an analogous post-conflict period, that between the World Wars (1919–1939), the Army found itself victorious, facing competing visions of the future with no obvious peer competitor in sight, Germany disarmed, and, with the League of Nations, an international collective security regime in place. The Army of that era struggled to arrive at a reasonable acquisition investment strategy for a host of reasons detailed below. Today's Army need not struggle. The tools and processes described in this monograph, if adopted and incorporated into the Army's current planning and programming practices, can help the Army plan the allocation of funds for those threats that are most dangerous and most imminent.

The Acquisition Environment After World War I

Following demobilization in 1919, the War Department had to assess the potential threats the country would be facing in the next 10 to 15 years. In 1919, the U.S.-Mexican border was beset by banditry and the spillover from the various factional conflicts that plagued post-revolutionary Mexico.[1] America's colonial possessions in the Far East

---

[1] See U.S. War Department (1920, pp. 244–245).

and Latin America required policing. There was a fear that Japan, fresh from acquiring some of Germany's imperial possessions in the Pacific, would soon be making a bid for hegemony in Asia.[2] And, of course, the ultimate long-term worry was that another large war could erupt in Europe if the Versailles settlement and the League of Nations regime lacked resolute international support. Unfortunately, little in the way of strategic guidance was provided by the administrations of the 1920s and early 1930s, and the service's internal intelligence assessment processes of the time were ill-equipped to handle the ambiguity and uncertainty that characterized the threat environment of 1919–1935.[3] Thus, by default, Army threat perceptions of the time were driven by near-term concerns and the personal judgments of senior officers. These practices, along with the very small acquisition budgets of the period, were to have serious consequences when more certain and substantial threats came into focus during the mid- to late 1930s.

Finally, in 1935 and 1936, after 15 years of uncertain threat priorities, the combination of Nazi Germany and Fascist Italy provided the Army with an enemy against which it had to develop a comprehensive investment strategy. Unfortunately, by this point the need to equip a larger U.S. Army was so imperative that a comprehensive investment strategy was impossible. What mattered was buying sufficient quantities of what was available from American industry and Army arsenals, even if that meant fielding forces equipped with previous-generation weapons.[4] These sudden procurement requirements necessitated large reductions in research and development activities that would have been part of a comprehensive investment strategy: a development that would have lingering effects into World War II.[5]

---

[2] See Miller (1991).

[3] Army threat analyses at the time went back and forth between emphasizing the threat of light guerrilla-type forces in the Western Hemisphere and that of potential medium- to heavyweight opponents in Europe and Asia.

[4] Johnson (1990, p. 264).

[5] The most famous examples of dated technology fielded by the U.S. Army through much of World War II being its tanks and anti-tank guns.

To be fair, the acquisition leadership during the interwar era was hobbled by a lack of funding, and the acquisition community itself was poorly organized. However, given the diverse visions of what the Army should become, the individual preferences of influential leaders such as Generals Pershing and MacArthur, and the divergent threat assessments of the time, it is doubtful that the acquisition system of the day could have produced an Army more suitable for the early engagements of World War II, even had the funding been available throughout the 1920s and 1930s. What was missing was a systematic way to assess, appraise, and harmonize acquisition decisions.

Today's Acquisition Environment

Today's acquisition community need not suffer from the pathologies that affected interwar acquisition decisions. In addition to enjoying richer resources than their predecessors, today's acquisition planners have the benefits of focused guidance all the way down the chain of command to the PEG level. They deploy an array of planning tools and processes designed to provide better acquisition decisions. For example, they employ the Automated Cost Estimating Integrated Tools (ACEIT) to simplify life cycle cost estimates; other tools are available through the Acquisition Information Management portal including the Web Army RDA Budget Update Computer System (WARBUCS). Nevertheless, there are analogies between today's acquisition environment and that of the interwar period.

- **Today's threat environment remains ambiguous.** Although the U.S. Army is currently involved in both a wide-ranging war against terrorism and a large, sustained commitment in Iraq, the threat of midsized powers equipped with nuclear weapons is growing more urgent and the potential for the emergence of a near-peer military competitor or a nuclear-armed regional power is not out of the question.
- **Balancing acquisition investments between near- and longer-term threats is as challenging now as it was in the 1920s and**

1930s. Eighty years ago, the Army was debating the effect that new technologies would have on warfighting.[6] Today's Army is likewise trying to understand the effect of new technologies on warfighting to devise sensible acquisition and development strategies to incorporate those technologies into its force structure.

These similarities are significant enough to encourage caution as the Army develops its acquisition program. This monograph, and the research behind it, is an attempt to provide the Army with a high-level acquisition planning tool that can help manage the uncertainties that complicate the acquisition planning, programming, and budgeting process. The AIM methodology introduced in this monograph is a planning tool that will help the Army make balanced, robust acquisition decisions that account for both near- and long-term threats and balances the materiel capabilities that the Army will need to address a wide spectrum of conflict. Absent such a tool, the Army acquisition system runs the risk of remaining in a post-conflict mindset.

---

[6] Internal combustion engines, aircraft, and wireless radio.

Assumption-Based Planning

This appendix describes the use of ABP to discover the Army's key, load-bearing assumptions and the process of testing them to assess their health.

ABP can be thought of as a series of questions used to interrogate an organization's plans and planning documents. These questions include:

- What are the key, load-bearing assumptions underlying the plan?
- What signs, if they appeared, would signal that the key assumptions were failing?
- On seeing these signposts, what actions might be taken to shape the future environment to prevent the assumptions from failing?
- In the event the shaping strategy fails, what hedging steps can be taken to help cope with the emergent circumstances?

Finding the Army's Key, Load-Bearing Assumptions

Typically, to unearth an organization's core assumptions, one can read its strategic plan, perform content analysis on its planning documents, and interview the organization's most senior leaders. The Army operates not from one strategic plan but from many, and its institutional thinking and judgments about the global security environment are informed by assessments from the intelligence community, directives from the Office of the Secretary of Defense, the plans and capabilities of its sister services, the capabilities delivered by the defense industrial sector, and the other departments within the executive branch of government. Given these many influences on Army planning, discovery of its key, load-bearing assumptions involved reading and data-mining a large number of documents.[1]

The 2004 Army Modernization Plan proved to be a key document. It reflected the threat and threat assumptions from the intelligence community. It also placed Army planning and assumptions in the context of Joint warfighting and unified action, reflecting the contributions of the rest of the U.S. military. The plan brought together expectations (assumptions) about technology and future operations and generally helped us manage the complexity of identifying key, load-bearing assumptions buried within a vast number of documents.

Searching for the Army's core assumptions involved conducting a key word search on the documents described above. The searches focused on the words "will," "must," and "needs" as clues to assumptions, along with a technique called "story-telling" (Dewar, 2002, p. 38) which helped us find declarative sentences and assertions that deserved attention as potential key assumptions. With the initial, raw assumptions in hand, we examined them for common underlying themes to determine whether the long list of initial assumptions could be condensed into a shorter and more fundamental set of assumptions. The initial set of assumptions were that

- Information superiority will be maintained.
- Data can be turned into actionable information very quickly.
- Precision maneuvers and precision fires can make up for mass.
- Very rapid strategic/operational deployability and sustainability are required and achievable.
- Technology can provide survivability to lighter weight and dispersed forces.

---

[1] These included threat estimates from Central Intelligence Agency (CIA); the Defense Intelligence Agency (DIA); the National Ground Intelligence Center (NGIC); the National Air and Space Intelligence Center (NASIC); the Army Science and Technology Master Plan; the Army Modernization Plan, FM 3-0, Operations; the FCS Operational Requirements Document (ORD); the FYDP; the Joint Strategic Capabilities Plan (JSCP); and Total Army Analysis documents.

- Units of action can be made very versatile.
- Enemies will span the range of capability but all will be competent and adaptive.
- A richer mix of capabilities will allow the Army to prevail against future foes.
- Integration of equipment, doctrine, training, infrastructure, and soldier/leader development will reveal and maximize new capabilities.

The next step was to vet these assumptions with the office of the Deputy for Acquisition and Systems Management. After several iterations of staffing, discussions with Army Acquisition officials, and our further attempts to condense and sharpen the assumptions still further, a list of six assumptions emerged:

- The U.S. Army will maintain operational and technological dominance over potential foes.
- The U.S. Army will require and maintain a capacity for rapid global deployment and self-sustainment in austere theaters.
- The U.S. Army will be increasingly adroit at managing complexity.
- U.S. Army budgets will sustain operational and technical dominance.
- The U.S. Army will rely on the capabilities of the Reserve Component and its sister services.
- The U.S. Army's enemies will span the range of capabilities but will all be competent and adaptive.

Are They Load-Bearing?

A key question about the list of assumptions is whether they are load-bearing: ABP jargon for the degree to which they really underpin and shape an organization's plans. To ascertain whether the revised assumptions were indeed load-bearing, we subjected them to "rationalization," which in ABP terms means to see how and whether all of the assumptions are connected to key features in the organization's plans. Given the abundance of planning documents, we tried the rationalization process against several Army core competencies, including the Objective Force Characteristics, as found in FM-1,[2] *The Army* (the Army's keystone manual listing the service's core functions), the Army Transformation Roadmap 2003, and the Acquisition Science and Technology Master Plan, which together describe the Army's plan to evolve for the future and exploit opportunities presented by scientific and technological breakthroughs. Overall, the "fit" of assumptions to plans was very good; no assumptions remained unconnected to key attributes of plans, and no planned actions were unconnected from any of the assumptions. As a result of the rationalization process, we concluded that these were, in fact, key, load-bearing assumptions, and that there were no other important, hidden assumptions influencing the Army's acquisition planning. Figure A.1, illustrates the rationalization process.

The arrows in the figure point from the assumptions to the Army core competencies they underpin. We built similar charts to trace the connectivity of its assumptions to key planning elements. From the rationalization process, we concluded that we had found the Army's key, load-bearing assumptions and that they were tightly connected to the Army's plans.

Are They Vulnerable?

Because these assumptions shape Army Acquisition investment planning and strategy, we sought to ascertain their health: Are they sound or becoming vulnerable because of advances in technology, changes in the international scene, or changes resulting from adjustments in U.S. policies? To answer the question, we conducted a number of brainstorming sessions to develop signposts of potential vulnerability:

---

[2] U.S. Army, Field Manual 1.

Figure A.1
**Rationalization: Aligning Assumptions with Key Planning Attributes in
FM-1, *The Army***

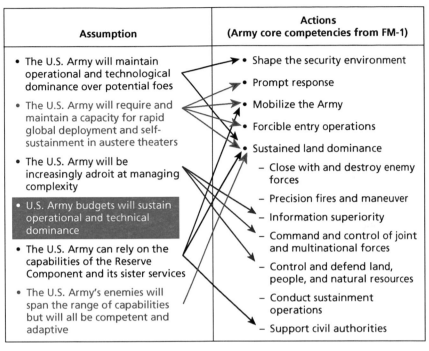

Assumption	Actions (Army core competencies from FM-1)
• The U.S. Army will maintain operational and technological dominance over potential foes	• Shape the security environment
	• Prompt response
• The U.S. Army will require and maintain a capacity for rapid global deployment and self-sustainment in austere theaters	• Mobilize the Army
	• Forcible entry operations
	• Sustained land dominance
• The U.S. Army will be increasingly adroit at managing complexity	– Close with and destroy enemy forces
	– Precision fires and maneuver
• U.S. Army budgets will sustain operational and technical dominance	– Information superiority
	– Command and control of joint and multinational forces
• The U.S. Army can rely on the capabilities of the Reserve Component and its sister services	– Control and defend land, people, and natural resources
• The U.S. Army's enemies will span the range of capabilities but will all be competent and adaptive	– Conduct sustainment operations
	– Support civil authorities

RAND *MG532-A.1*

phenomena that, if observed and left unchecked, would render a key, load-bearing assumption vulnerable. Moreover, as noted in the introduction to this monograph, ABP calls for shaping actions to keep core assumptions healthy and hedging actions to cope with increasing and unstoppable vulnerability. The paragraphs below describe the signposts of vulnerability that emerged from our brainstorming sessions, followed by descriptions of acquisition-oriented shaping and hedging for each assumption.

- **The U.S. Army will maintain operational and technological dominance over potential foes.**

Signposts of potential vulnerability:

a. *Potential peer adversaries (China, a Unified Europe, Japan, Russia) significantly intensify their investments in one or more of the following:*

i. Offensive and defensive information technology for the military

Shaping actions:

1. Invest in force protection procurement (information, communications, networks)
2. Invest in command and control procurement
3. Invest in RSTA procurement
4. Invest in remote support procurement (computer network attack capabilities)
5. Invest in force protection R&D (information, communications, networks)
6. Invest in command and control R&D
7. Invest in RSTA R&D

Hedging actions:

1. Increase investment in force protection procurement (information, communications, networks)
2. Increase investment in remote support procurement (training systems)
3. Increase investment in command and control procurement (redundant systems)
4. Increase investment in RSTA procurement (redundant systems)
5. Increase investment in force protection R&D
6. Increase investment in command and control R&D
7. Increase investment in RSTA R&D
8. Invest in systems integration R&D (modeling and simulation for information technology analysis)

ii. Air superiority capabilities that are equivalent to U.S. capabilities

Shaping actions:

1. Invest in force protection procurement (next-generation air defense (AD) systems)
2. Invest in force protection R&D (AD systems)

Hedging actions:

1. Increase investment in force protection R&D (AD systems)
2. Increase investment in force protection procurement (next-generation AD systems)
3. Recapitalize and modernize force protection systems (AD systems)
4. Increase remote support procurement (AD training systems)

iii. Naval and anti-naval capabilities that threaten U.S. Navy command of the oceans

Shaping actions:

1. Invest in forward support equipment procurement (forward deployment equipment)
2. Invest in remote support procurement (deployment infrastructure: rails, ports, roads)
3. Invest in force protection procurement (port security)

Hedging actions:

1. Invest in close battle R&D (develop air-transportable, lethal, survivable, air-supportable close battle systems)
2. Invest in mobility R&D (develop air-transportable, reliable, air-supportable mobility systems)
3. Invest in direct fire R&D (develop air-transportable, lethal, survivable, air-supportable direct fire systems)

4. Invest in indirect fire R&D (develop air-transportable, lethal, survivable, air-supportable indirect fire systems)
5. Invest in forward support R&D (develop air-transportable, air-supportable forward support systems)
6. Recapitalize and modernize close battle (placement as prepositioned materiel)
7. Recapitalize and modernize mobility (placement as prepositioned materiel)
8. Recapitalize and modernize direct fire (placement as prepositioned materiel)
9. Recapitalize and modernize indirect fire (placement as prepositioned materiel)
10. Recapitalize and modernize forward support (placement as prepositioned materiel)
11. Invest in procurement of remote support (training resources for forced entry operations)

iv. Long-range strategic transport

Shaping actions: None for the acquisition community.

Hedging actions:

1. Invest in close battle R&D (develop air-transportable, lethal, survivable, air-supportable close battle systems)
2. Invest in mobility R&D (develop air-transportable, reliable, air-supportable mobility systems)
3. Invest in direct fire R&D (develop air-transportable, lethal, survivable, air-supportable direct fire systems)
4. Invest in indirect fire R&D (develop air-transportable, lethal, survivable, air-supportable indirect fire systems)
5. Invest in forward support R&D (develop air-transportable, air-supportable forward support systems)
6. Recapitalize and modernize close battle (placement as prepositioned materiel)

7. Recapitalize and modernize mobility (placement as prepositioned materiel)
8. Recapitalize and modernize direct fire (placement as prepositioned materiel)
9. Recapitalize and modernize indirect fire (placement as prepositioned materiel)
10. Recapitalize and modernize forward support (placement as prepositioned materiel)
11. Invest in procurement of remote support (training resources for forced entry operations)

v. Military research and development

Shaping actions:

1. Invest in close battle R&D (develop next-generation systems and equipment, explore longer-range possibilities)
2. Invest in mobility R&D (develop next-generation systems and equipment, explore longer-range possibilities)
3. Invest in direct fire R&D (develop next-generation systems and equipment, explore longer-range possibilities)
4. Invest in indirect fire R&D (develop next-generation systems and equipment, explore longer-range possibilities)
5. Invest in forward support R&D (develop next-generation systems and equipment, explore longer-range possibilities)
6. Invest in force protection R&D (develop next-generation systems and equipment, explore longer-range possibilities)
7. Invest in remote support R&D (develop next-generation systems and equipment, explore longer-range possibilities; emphasize training systems to maintain the U.S. Army's current training dominance)
8. Invest in command and control R&D (develop next-generation systems and equipment, explore longer-range possibilities)
9. Invest in RSTA R&D (develop next-generation systems and equipment, explore longer-range possibilities)

10. Invest in systems integration R&D (develop next-generation systems and equipment, explore longer-range possibilities)

Hedging actions:

1. Increase investment in close battle R&D (develop leap-ahead capabilities)
2. Increase investment in mobility R&D (develop leap-ahead capabilities)
3. Increase investment in direct fire system R&D (develop leap-ahead capabilities)
4. Increase investment in indirect fire R&D (develop leap-ahead capabilities)
5. Increase investment in forward support R&D (develop leap-ahead capabilities)
6. Increase investment in force protection R&D (develop leap-ahead capabilities)
7. Increase investment in remote support R&D (develop leap-ahead capabilities)
8. Increase investment in command and control R&D (develop leap-ahead capabilities)
9. Increase investment in RSTA R&D (develop leap-ahead capabilities)
10. Increase investment in systems integration R&D (develop leap-ahead capabilities)

b. *Potential regional adversaries do two things: demonstrate an intention to dominate a region militarily; demonstrate a growing military capability through investment in anti-access capabilities that either (or both) prevent or hamper U.S. military entry into the region or increase defense spending on offensive and regional power projection capabilities.*

Shaping actions: None for the acquisition community.

Hedging actions:

1. Invest in close battle procurement (prepositioning of systems and equipment)
2. Increase investment in mobility procurement (prepositioning of systems and equipment for forced entry)
3. Invest in direct fire procurement (prepositioning of systems and equipment)
4. Increase investment in indirect fire procurement (prepositioning of systems and equipment for forced entry, e.g., long-range precision)
5. Invest in forward support procurement (prepositioning of systems and equipment)
6. Increase investment in force protection procurement (prepositioning of systems and equipment for forced entry)
7. Significantly increase investment in remote support procurement (buy medium/fast transport systems, preposition systems and equipment in forward theaters, adjust the mix of air, sea, ground transport, and fund training systems for forced entry)
8. Invest in command and control procurement (prepositioning of systems and equipment)
9. Increase investment in RSTA procurement (prepositioning of systems and equipment for forced entry)

c. *Nuclear weapons and delivery systems appear in formerly non-nuclear states.*

Shaping actions: None for the acquisition community.

Hedging actions:

1. Invest in mobility procurement (prepositioning of systems and equipment for forced entry)
2. Increase investment in indirect fire procurement (prepositioning of systems and equipment for forced entry, e.g., long-range precision)
3. Invest in force protection procurement (prepositioning of systems and equipment for forced entry)

4. Significantly increase investment in command and control procurement (hardening of equipment, redundancy, support of distributed operations)
5. Significantly increase investment in RSTA procurement (hardening of equipment, identification of adversary WMD systems, support of distributed operations and to support forced entry operations)
6. Invest in close battle R&D (develop air-transportable, lethal, survivable, air-supportable close battle systems)
7. Invest in mobility R&D (develop air-transportable, reliable, air-supportable mobility systems)
8. Invest in direct fire R&D (develop air-transportable, lethal, survivable, air-supportable direct fire systems)
9. Increase investment in indirect fire R&D (develop air-transportable, lethal, survivable, air-supportable direct fire systems and for systems intended for use on hardened sites, e.g., deep penetration, electromagnetic pulse (EMP), tactical nuclear)
10. Invest in forward support R&D (develop air-transportable, lethal, survivable, air-supportable direct fire systems)
11. Increase investment in remote support R&D (medical research for pretreatment and treatment of injuries associated with nuclear blasts and for training systems for operating in contaminated environments)

d. *Key defense industries in the United States fail, stagnate, or consolidate or move overseas.*

Shaping actions:

1. Invest in remote support procurement (maintenance of a healthy industrial base, e.g., contract for surge potential, fund independent research and development (IRD), develop dual use facilities)
2. Invest in remote support R&D (industry and academia R&D in key fields)

Hedging actions:

1. Invest in close battle procurement (maintain large war reserve stockpiles of ammunition)
2. Invest in direct fire procurement (maintain large war reserve stockpiles of ammunition)
3. Invest in indirect fire procurement (maintain large war reserve stockpiles of ammunition)
4. Invest in forward support procurement (maintain large war reserve stockpiles of fuel, food, spare-parts, batteries)
5. Invest in force protection procurement (maintain large war reserve stockpiles of overgarments, decontamination equipment, etc.)
6. Significantly increase investment in remote support procurement (identify and subsidize strategic industries, establish business relationships with off-shore suppliers and firms, develop additional government-owned production capability for important, militarily unique materiel)
7. Increase close battle R&D (maintain knowledge)
8. Increase mobility R&D (maintain knowledge)
9. Increase direct fire R&D (maintain knowledge)
10. Increase indirect fire R&D (maintain knowledge)
11. Increase forward support R&D (maintain knowledge)
12. Increase force protection (maintain knowledge)
13. Increase command and control R&D (maintain knowledge)
14. Increase RSTA R&D (maintain knowledge)
15. Increase systems integration R&D (maintain knowledge)

e. *Countermeasures to important U.S. military technologies (stealth, networks, communications, etc.) appear and begin proliferating.*

Shaping actions:

1. Invest in close battle R&D (prevent surprises, stay ahead of potential adversaries)

2. Invest in mobility R&D (prevent surprises, stay ahead of potential adversaries)
3. Invest in direct fire R&D (prevent surprises, stay ahead of potential adversaries)
4. Invest in indirect fire R&D (prevent surprises, stay ahead of potential adversaries)
5. Invest in forward support R&D (prevent surprises, stay ahead of potential adversaries)
6. Invest in force protection R&D (prevent surprises, stay ahead of potential adversaries)
7. Invest in remote support R&D (prevent surprises, stay ahead of potential adversaries)
8. Invest in command and control R&D (prevent surprises, stay ahead of potential adversaries)
9. Invest in RSTA R&D (prevent surprises, stay ahead of potential adversaries)
10. Invest in systems integration R&D (prevent surprises, stay ahead of potential adversaries)

Hedging actions:

1. Invest in close battle procurement (systems to substitute mass for technology)
2. Invest in mobility procurement (systems to substitute mass for technology)
3. Invest in direct fire procurement (systems to substitute mass for technology)
4. Invest in indirect fire procurement (systems to substitute mass for technology)
5. Invest in remote support procurement (maintain U.S. training superiority)
6. Invest in command and control procurement (redundant capability)
7. Invest in RSTA procurement (redundant capability)
8. Increase investment in close battle R&D (prevent surprises, stay ahead of potential adversaries)

9. Increase investment in mobility R&D (prevent surprises, stay ahead of potential adversaries)
10. Increase investment in direct fire R&D (prevent surprises, stay ahead of potential adversaries)
11. Increase investment in indirect fire R&D (prevent surprises, stay ahead of potential adversaries)
12. Increase investment in forward support R&D (prevent surprises, stay ahead of potential adversaries)
13. Increase investment in force protection R&D (prevent surprises, stay ahead of potential adversaries)
14. Increase investment in command and control R&D (prevent surprises, stay ahead of potential adversaries)
15. Increase investment in RSTA R&D (prevent surprises, stay ahead of potential adversaries)
16. Increase investment in systems integration R&D (prevent surprises, stay ahead of potential adversaries)

f. *New weapons or doctrine based on new technologies appear outside the U.S. military.*

Shaping actions:

1. Invest in close battle R&D (prevent surprises, stay ahead of potential adversaries)
2. Invest in mobility R&D (prevent surprises, stay ahead of potential adversaries)
3. Invest in direct fire R&D (prevent surprises, stay ahead of potential adversaries)
4. Invest in indirect fire R&D (prevent surprises, stay ahead of potential adversaries)
5. Invest in forward support R&D (prevent surprises, stay ahead of potential adversaries)
6. Invest in force protection R&D (prevent surprises, stay ahead of potential adversaries)
7. Invest in remote support R&D (prevent surprises, stay ahead of potential adversaries)

8. Invest in command and control R&D (prevent surprises, stay ahead of potential adversaries)
9. Invest in RSTA R&D (prevent surprises, stay ahead of potential adversaries)
10. Invest in systems integration R&D (prevent surprises, stay ahead of potential adversaries)

Hedging actions:

1. Increase investment in close battle R&D (prevent surprises, stay ahead of potential adversaries)
2. Increase investment in mobility R&D (prevent surprises, stay ahead of potential adversaries)
3. Increase investment in direct fire R&D (prevent surprises, stay ahead of potential adversaries)
4. Increase investment in indirect fire R&D (prevent surprises, stay ahead of potential adversaries)
5. Increase investment in forward support R&D (prevent surprises, stay ahead of potential adversaries)
6. Increase investment in force protection R&D (prevent surprises, stay ahead of potential adversaries)
7. Increase investment in remote support R&D (prevent surprises, stay ahead of potential adversaries)
8. Increase investment in command and control R&D (prevent surprises, stay ahead of potential adversaries)
9. Increase investment in RSTA R&D (prevent surprises, stay ahead of potential adversaries)
10. Increase investment in systems integration R&D (prevent surprises, stay ahead of potential adversaries)

- **The U.S. Army will require and maintain a capacity for rapid global deployment and self-sustainment in austere theaters.**

Signposts of potential vulnerability:

a. *Potential adversaries invest in/deploy significant anti-access capability.*

Shaping actions:

1. Invest in remote support procurement (fast sealift)
2. Invest in RSTA procurement

Hedging actions:

1. Invest in close battle procurement (equipment and systems for prepositioning in regions of likely conflict)
2. Increase investment in mobility procurement (equipment and systems for prepositioning in regions of likely conflict for forced entry operations)
3. Invest in direct fire procurement (equipment and systems for prepositioning in regions of likely conflict)
4. Increase investment in indirect fire procurement (equipment and systems for prepositioning in regions of likely conflict for forced entry operations)
5. Invest in forward support procurement (equipment and systems for prepositioning in regions of likely conflict)
6. Increase investment in force protection procurement (equipment and systems for prepositioning in regions of likely conflict for forced entry operations)
7. Increase investment in remote support procurement (diversify strategic lift; adjust mix of air, sea, ground transport, and training equipment for forced entry operations)
8. Invest in command and control procurement (equipment and systems for prepositioning in regions of likely conflict)
9. Increase investment in RSTA procurement (equipment and systems for prepositioning in regions of likely conflict for forced entry operations)
10. Invest in close battle R&D (develop air-transportable, lethal, survivable, air-supportable close battle systems)

11. Invest in mobility R&D (develop air-transportable, reliable, air-supportable mobility systems)
12. Invest in direct fire R&D (develop air-transportable, lethal, survivable, air-supportable direct fire systems)
13. Invest in indirect fire R&D (develop air-transportable, lethal, survivable, air-supportable indirect fire systems)
14. Invest in forward support R&D (develop air-transportable, air-supportable indirect fire systems)
15. Invest in remote support R&D (improve training in complex operations)

b. *Host nation support (HNS) becomes questionable or contingent on nature and location of U.S. operations.*

Shaping actions: None for the acquisition community.

Hedging actions:

1. Increase investment in close battle R&D (systems that are more efficient and reliable)

2. Increase investment in mobility R&D (systems that are more efficient and reliable)
3. Increase investment in direct fire R&D (systems that are more efficient and reliable)
4. Increase investment in indirect fire R&D (systems that are more efficient and reliable)
5. Increase investment in forward support R&D (systems that are more efficient and reliable)
6. Increase investment in force protection R&D (systems that are more efficient and reliable)

c. U.S. foreign policy becomes isolationist.

Shaping actions: None for the acquisition community.

Hedging actions:

1. Increase close battle R&D (retain military know-how)
2. Increase mobility R&D (retain military know-how)
3. Increase direct fire R&D (retain military know-how)
4. Increase indirect fire R&D (retain military know-how)
5. Increase command and control R&D (retain military know-how)
6. Increase RSTA R&D (retain military know-how)
7. Increase systems integration R&D (retain military know-how)

d. *Very significant terrorist activity inside the United States results in re-missioning of military forces to support anti-terrorist activities inside the United States.*

Shaping actions:

1. Increase investment in close battle procurement (develop systems and equipment to maintain the fight against terrorists overseas)
2. Increase investment in mobility procurement (develop systems and equipment to maintain the fight against terrorists overseas)
3. Increase investment in RSTA procurement (develop systems and equipment to maintain the fight against terrorists overseas)
4. Increase investment in systems integration R&D (maintain the fight against terrorists overseas)

Hedging actions:

1. Invest in command and control procurement (materiel for coordinated cooperation with law enforcement, e.g., Rear Area Operations Center (RAOC) and Civil-Military Operations Center (CMOC)
2. Increase investment in close combat procurement equipment (useful in both domestic anti-terrorist work and higher-end combat operations)
3. Increase investment in mobility procurement equipment (useful in both domestic anti-terrorist work and higher-end combat operations, e.g., Stryker, aviation)

4. Increase investment in RSTA procurement (e.g., ISR for the soldier, instrumenting cities—deploying active and passive sensors throughout the landscape that provide input to ISR/RSTA systems improving soldiers' awareness of enemy activities in the area)

5. Increase in command and control R&D (multiagency)

6. Increase in RSTA R&D (for urban environments)

e. *Technically feasible solutions for rapid deployment of U.S. Army forces are unavailable or unaffordable.*

Shaping actions:

1. Invest in close battle R&D (develop air-transportable, lethal, survivable, air-supportable close battle systems)

2. Invest in mobility R&D (develop air transportable, reliable, air-supportable mobility systems)

3. Invest in direct fire R&D (develop air-transportable, lethal, survivable, air-supportable direct fire systems)

4. Invest in indirect fire R&D (develop air-transportable, lethal, survivable, air-supportable indirect fire systems)

5. Invest in forward support R&D (develop air-transportable, air-supportable forward support systems)

6. Invest in force protection R&D (develop air-transportable, air-supportable force protection systems)

7. Invest in RSTA R&D (develop air-transportable, air-supportable RSTA systems)

Hedging actions:

1. Invest in close battle procurement (for prepositioning of systems and equipment)

2. Invest in mobility procurement (for prepositioning of systems and equipment)

3. Invest in direct fire procurement (for prepositioning of systems and equipment)

4. Invest in indirect fire procurement (for prepositioning of systems and equipment)
5. Invest in forward support procurement (for prepositioning of systems and equipment)
6. Invest in force protection procurement (for prepositioning of systems and equipment)
7. Increase investment in remote support procurement (for prepositioning of systems and equipment and for additional roll-on/roll-off (ROROs) cargo ships, high-speed sealift)
8. Invest in command and procurement (for prepositioning of control systems and equipment)
9. Invest in RSTA procurement (for prepositioning of systems and equipment)
10. Re-capitalize and modernize close battle systems
11. Re-capitalize and modernize mobility systems
12. Re-capitalize and modernize direct fire systems
13. Re-capitalize and modernize indirect fire systems
14. Re-capitalize and modernize forward support systems
15. Re-capitalize and modernize protection systems
16. Re-capitalize and modernize remote support systems
17. Re-capitalize and modernize command and control systems
18. Re-capitalize and modernize RSTA systems

f. *Other services or OSD dispute that the Army has a requirement for rapid global maneuver or OSD fails to resource that requirement.*

Shaping actions:

1. Increase investment in remote support R&D (e.g., multi-service Deployability Advanced Concept Technology Demonstrations (ACTDs))
2. Invest in close battle R&D (develop air-transportable, lethal, survivable, air-supportable close battle systems)
3. Invest in mobility R&D (develop air-transportable, reliable, air-supportable mobility systems)

4. Invest in direct fire R&D (develop air-transportable, lethal, survivable, air-supportable direct fire systems)
5. Invest in indirect fire R&D (develop air-transportable, lethal, survivable, air-supportable indirect fire systems)
6. Invest in forward support R&D (develop air-transportable, air-supportable forward support systems)
7. Invest in force protection R&D (develop air-transportable, air-supportable force protection systems)
8. Invest in RSTA R&D (develop air-transportable, air-supportable RSTA systems)

Hedging actions:

1. Invest in close battle procurement (for prepositioning of systems and equipment)
2. Invest in mobility procurement (for prepositioning of systems and equipment)
3. Invest in direct fire procurement (for prepositioning of systems and equipment)
4. Invest in indirect fire procurement (for prepositioning of systems and equipment)
5. Invest in forward support procurement (for prepositioning of systems and equipment)
6. Invest in force procurement (for prepositioning of protection systems and equipment)
7. Invest in remote support procurement (for prepositioning of systems and equipment and deployment systems)
8. Invest in command and control procurement (for prepositioning of systems and equipment)
9. Invest in RSTA procurement (for prepositioning of systems and equipment)

g. *U.S. interest in austere theaters wanes.*

Shaping actions: None for the acquisition community.

Hedging actions:

1. Reduce investment in remote support procurement
2. Reduce investment in forward support procurement

- **The U.S. Army will be increasingly adroit at managing complexity.**

Signposts of potential vulnerability:

a. *Continued and very high OPTEMPO severely constrains training opportunities for most Army units and limits the availability of test units to support experimentation.*

Shaping actions:

1. Increase investment in remote support (training technologies, e.g., distance learning and other technologies to enable training during deployments)
2. Invest in systems integration R&D (modeling and simulation, automation technology)
3. Invest in command and control procurement (e.g., long-distance collaboration tools)

Hedging actions: None the acquisition community can take.

b. *Emergence of very sophisticated enemies who are able to purposely inject added complexity while simultaneously reducing the U.S. Army's ability to deal with complexity.*

Shaping actions:

1. Increase investment in remote support procurement (training systems to improve the ability of soldiers to manage complexity)
2. Invest in command and control procurement
3. Invest in RSTA procurement

4. Increase investment in remote support R&D (training systems to improve the ability of soldiers to manage complexity)
5. Invest in command and control R&D
6. Invest in RSTA R&D
7. Increase investment in systems integration R&D

Hedging actions:

1. Invest in close battle procurement (equipment and systems to add mass to U.S. forces)
2. Invest in mobility procurement (equipment and systems to add mass to U.S. forces)
3. Invest in direct fire procurement (equipment and systems to add mass to U.S. forces)
4. Invest in indirect fire procurement (equipment and systems to add mass to U.S. forces)
5. Invest in forward support procurement (equipment and systems to add mass to U.S. forces)
6. Increase investment in remote support procurement (equipment and systems to add mass to U.S. forces and in training systems to improve the ability of soldiers to manage complexity)
7. Increase investment in command and control procurement (to help deal with complex operations and to provide redundant systems to U.S. forces)
8. Increase investment in RSTA procurement (to help deal with complex operations and to provide redundant systems to U.S. forces)
9. Increase investment in remote support R&D (training systems to improve the ability of soldiers to manage complexity)
10. Increase investment in command and control R&D
11. Increase investments in RSTA R&D

c. *The Army cannot recruit the right kind of people.*

Shaping actions:

1. Increase investment in remote support procurement (training systems to improve the ability of marginal soldiers to manage complexity)
2. Increase investment in remote support R&D (training systems to improve the ability of marginal soldiers to manage complexity)

Hedging actions:

1. Invest in mobility procurement (automated systems to replace soldiers with technology, if currently available)
2. Invest in direct fire procurement (automated systems to replace soldiers with technology, if currently available)
3. Invest in indirect fire procurement (automated systems to replace soldiers with technology, if currently available)
4. Invest in forward support procurement (automated systems to replace soldiers with technology, if currently available)
5. Increase investment in remote support procurement (automated systems to replace soldiers with technology, if currently available, and training technology and systems)
6. Invest in command and control procurement (automated systems to replace soldiers with technology, if currently available)
7. Increase investment in remote support R&D (training technology and systems)
8. Increase investment in systems integration R&D (automation technology)

d. *Numerous operational, planning, logistic, and intelligence failures by the Army occur over a short period of time.*

Shaping actions:

1. Invest in forward support procurement
2. Invest in remote support procurement
3. Invest in command and control procurement
4. Invest in RSTA procurement
5. Invest in forward support R&D
6. Invest in remote support R&D

7. Invest in command and control R&D
8. Invest in RSTA R&D
9. Increase investment in systems integration R&D

Hedging actions:

1. Increase investment in remote support procurement (automated logistic systems and training systems to train soldiers and leaders to plan and function well)
2. Increase investment in command and control procurement
3. Invest in remote support R&D (continually improve soldier and leader capability)
4. Increase investment in command and control R&D
5. Recapitalize and modernize close battle
6. Recapitalize and modernize mobility
7. Recapitalize and modernize direct fire
8. Recapitalize and modernize indirect fire
9. Recapitalize and modernize forward support
10. Recapitalize and modernize force protection

e. *High-tech units fare poorly at the National Training Center (NTC)/Joint Readiness Training Center (JRTC) and do not improve over time.*

Shaping actions:

1. Invest in indirect fire procurement (precision munitions)
2. Invest in remote support procurement (realistic and challenging training for high-tech units)
3. Invest in command and control procurement
4. Invest in RSTA procurement
5. Invest in indirect fire systems R&D (of precision munitions)
6. Invest in remote support R&D (improve the ability of soldiers and leaders to adapt to high-tech units)
7. Invest in command and control R&D
8. Invest in RSTA R&D

Hedging actions:

1. Increase investment in systems integration R&D
2. Increase investment in command and control R&D of technology and systems
3. Increase investment in procurement of command and control technology and systems
4. Recapitalize and modernize close battle
5. Recapitalize and modernize mobility
6. Recapitalize and modernize direct fire
7. Recapitalize and modernize indirect fire
8. Recapitalize and modernize forward support
9. Recapitalize and modernize force protection

f. *Technology Readiness Levels (TRL) of critical technologies mature much more slowly than expected.*

Shaping actions:

1. Invest in close battle R&D
2. Invest in mobility R&D
3. Invest in direct fire R&D
4. Invest in indirect fire R&D
5. Invest in forward support R&D
6. Invest in force protection R&D
7. Invest in remote support R&D
8. Invest in command and control R&D
9. Invest in RSTA R&D
10. Invest in systems integration R&D

Hedging actions:

1. Invest in interim close battle procurement
2. Invest in interim mobility procurement
3. Invest in interim direct fire procurement
4. Invest in interim indirect fire procurement
5. Invest in interim forward support procurement
6. Invest in interim force protection procurement
7. Invest in command and control procurement (interim systems)

8. Invest in RSTA procurement (interim systems)
9. Increase investments in close battle R&D
10. Increase investments in mobility R&D
11. Increase investments in direct fire R&D
12. Increase investments in indirect fire R&D
13. Increase investments in forward support R&D
14. Increase investments in force protection R&D
15. Increase investments in remote support R&D
16. Increase investments in command and control R&D
17. Increase investments in RSTA R&D in systems and equipment
18. Increase investments in systems integration R&D
19. Recapitalize and modernize close battle
20. Recapitalize and modernize mobility
21. Recapitalize and modernize direct fire
22. Recapitalize and modernize indirect fire
23. Recapitalize and modernize forward support
24. Recapitalize and modernize force protection

g. *Systems integration proves difficult.*

Shaping actions:

1. Invest in systems integration R&D

Hedging actions:

1. Increase investment in close battle R&D
2. Increase investment in direct fire R&D
3. Increase investment in indirect fire R&D
4. Increase investment in force protection R&D
5. Increase investment in command and control R&D
6. Increase investment in RSTA R&D
7. Increase investment in systems integration R&D
8. Recapitalize and modernize close battle
9. Recapitalize and modernize mobility
10. Recapitalize and modernize direct fire

11. Recapitalize and modernize indirect fire
12. Recapitalize and modernize forward support
13. Recapitalize and modernize force protection

- **U.S. Army budgets will sustain operational and technical dominance.**

Signposts of potential vulnerability:

a. *Congressional support for large defense budgets wanes with respect to other budget priorities.*

Shaping actions: None for the acquisition community.

Hedging actions:

1. Invest in close battle R&D (stockpile technology)
2. Invest in mobility R&D (stockpile technology)
3. Invest in direct fire R&D (stockpile technology)
4. Invest in indirect fire R&D (stockpile technology)
5. Invest in forward support R&D (stockpile technology)
6. Invest in force protection R&D (stockpile technology)
7. Invest in remote support R&D (stockpile technology)
8. Invest in command and control R&D (stockpile technology)
9. Invest in RSTA R&D (stockpile technology)
10. Invest in systems integration R&D (stockpile technology)

b. *OPTEMPO consumes a large percentage of the defense budget over time.*

Shaping actions: None from the acquisition community.

Hedging actions:

1. Recapitalize and modernize close battle
2. Recapitalize and modernize mobility

3. Recapitalize and modernize direct fire
4. Recapitalize and modernize fire
5. Recapitalize and modernize forward support
6. Recapitalize and modernize force protection
7. Recapitalize and modernize remote support
8. Recapitalize and modernize command and control
9. Recapitalize and modernize RSTA

c. *DoD-level decisions concerning missions reallocate resources in favor of other services.*

Shaping actions: None for the acquisition community.

Hedging actions:

1. Invest in remote support procurement (align Army capabilities with OSD priorities)
2. Increase investment in close battle R&D (align Army capabilities with OSD priorities)
3. Increase investment in mobility R&D (align Army capabilities with OSD priorities)
4. Increase investment in direct fire R&D (align Army capabilities with OSD priorities)
5. Increase investment in indirect fire R&D (align Army capabilities with OSD priorities)
6. Increase investment in forward support R&D (align Army capabilities with OSD priorities)
7. Increase investment in force protection R&D (align Army capabilities with OSD priorities)
8. Increase investment in remote support R&D (align Army capabilities with OSD priorities)
9. Increase investment in command and control R&D (align Army capabilities with OSD priorities)
10. Increase investment in RSTA R&D (align Army capabilities with OSD priorities)

11. Increase investment in systems integration R&D (align Army capabilities with OSD priorities)

d. *Personnel costs continue to increase significantly.*

Shaping actions:

1. Increase investment in indirect fire procurement (replace manpower with technology)
2. Increase investment in forward support procurement (replace manpower with technology)
3. Increase investment in remote support procurement (replace manpower with technology)
4. Increase investment in command and control procurement (replace manpower with technology)
5. Increase investment in RSTA procurement (replace manpower with technology)
6. Increase investment in close battle R&D (replace manpower with technology)
7. Increase investment in mobility R&D (replace manpower with technology)
8. Increase investment in direct fire R&D (replace manpower with technology)
9. Increase investment in indirect fire R&D (replace manpower with technology)
10. Increase investment in forward support R&D (replace manpower with technology)
11. Increase investment in force protection R&D (replace manpower with technology)
12. Increase investment in remote support R&D (replace manpower with technology)
13. Increase investment in command and control R&D (replace manpower with technology)
14. Increase investment in RSTA R&D (replace manpower with technology)

15. Increase investment in systems integration R&D (replace manpower with technology)

Hedging actions:

1. Significantly increase close battle R&D (automated systems to replace soldiers with technology)
2. Significantly increase mobility R&D (automated systems to replace soldiers with technology)
3. Significantly increase direct fire R&D (automated systems to replace soldiers with technology)
4. Significantly increase indirect fire R&D (automated systems to replace soldiers with technology)
5. Significantly increase forward support R&D (automated systems to replace soldiers with technology)
6. Significantly increase force protection R&D (automated systems to replace soldiers with technology)
7. Significantly increase remote support R&D (automated systems to replace soldiers with technology and to improve training)
8. Significantly increase command and control R&D (automated systems to replace soldiers with technology)
9. Significantly increase RSTA R&D (automated systems to replace soldiers with technology)
10. Significantly increase systems integration R&D (automated systems to replace soldiers with technology)

- **The U.S. Army will rely on the capabilities of the Reserve Component and its sister services.**

Signposts of potential vulnerability:

a. *OPTEMPO challenges the ability of reserves to train.*

Shaping actions:

1. Increase investment in remote support procurement (distance learning technologies, virtual reality technology, gaming and simulation technology to maximize available training opportunities)

Hedging actions:

1. Increase investment in indirect close battle R&D (replace manpower with technology)
2. Increase investment in mobility R&D (replace manpower with technology)
3. Increase investment in direct fire R&D (replace manpower with technology)
4. Increase investment in indirect fire R&D (replace manpower with technology)
5. Increase investment in forward support R&D (replace manpower with technology)
6. Increase investment in force protection R&D (replace manpower with technology)
7. Increase investment in remote support R&D (replace manpower with technology and training to get more out of each soldier and leader)
8. Increase investment in command and control R&D (replace manpower with technology)
9. Increase investment in RSTA R&D (replace manpower with technology)
10. Increase investment in systems integration R&D (replace manpower with technology)

b. *Multiple contingencies divert the attention of the Reserve Component and Army sister services from Army priorities.*

Shaping actions: None for the acquisition community.

Hedging actions:

1. Increase investment in indirect fire procurement
2. Increase investment in forward support procurement
3. Increase investment in remote support procurement
4. Increase investment in RSTA procurement

c. *Sister services allocate their resources in ways that do not support Army priorities.*

Shaping actions: None for the acquisition community.

Hedging actions:

1. Increase investment in close battle R&D (align Army capabilities with OSD priorities)
2. Increase investment in mobility R&D (align Army capabilities with OSD priorities)
3. Increase investment in direct fire R&D (align Army capabilities with OSD priorities)
4. Increase investment in indirect fire R&D (align Army capabilities with OSD priorities)
5. Increase investment in forward support R&D (align Army capabilities with OSD priorities)
6. Increase investment in force protection R&D (align Army capabilities with OSD priorities)
7. Increase investment in remote support R&D (align Army capabilities with OSD priorities)
8. Increase investment in command and control R&D (align Army capabilities with OSD priorities) and RSTA R&D (align Army capabilities with OSD priorities)
9. Increase investment in systems integration R&D (align Army capabilities with OSD priorities)

d. *Other services and the Reserve Component are unable to recruit and retain sufficient numbers of the right kind of people.*

Shaping actions: None for the acquisition community.

Hedging actions:

1. Increase investment in indirect close battle R&D (replace manpower with technology)
2. Increase investment in mobility R&D (replace manpower with technology)
3. Increase investment in direct fire R&D (replace manpower with technology)
4. Increase investment in indirect fire R&D (replace manpower with technology)
5. Increase investment in forward support R&D (replace manpower with technology)
6. Increase investment in force protection R&D (replace manpower with technology)
7. Increase investment in remote support R&D (replace manpower with technology)
8. Increase investment in command and control R&D (replace manpower with technology)
9. Increase investment in RSTA R&D (replace manpower with technology)
10. Increase investment in systems integration R&D (replace manpower with technology)

e. *Force structure reductions reduce the number of contingencies other services can manage.*

Shaping actions: None for the acquisition community.

Hedging actions:

1. Increase investment in indirect fire procurement
2. Increase investment in forward support procurement
3. Increase investment in force protection procurement
4. Increase investment in remote support procurement
5. Increase investment in RSTA procurement

Using unclassified sources only, we assessed whether the signposts identified above are emerging. Figure A.2 is the result of that analysis and indicates where we believe that signposts have begun to emerge. The signpost analysis here, though, is necessarily cursory and was meant only to provide input to AIM.[3] If AIM is adopted into the PPBS, this important analysis will require detailed assessments by experts with access to classified and unclassified information. In addition, determining the emergent characteristics of signposts is not trivial. Most important, the description of the signposts must be precise enough to allow assessment of their emergent characteristics. Qualifying words such as "significant" are useful. Comparative statements also allow an analyst to make good assessments of signpost emergence, e.g., "Country A spends twice as much on defense as Country B." Finally, any analysis of signpost emergence must be accomplished iteratively. During the relatively cursory signpost analysis that informed this exercise of AIM, we found that it had a tendency to inflate the significance of the

Figure A.2
Signposts of Vulnerability

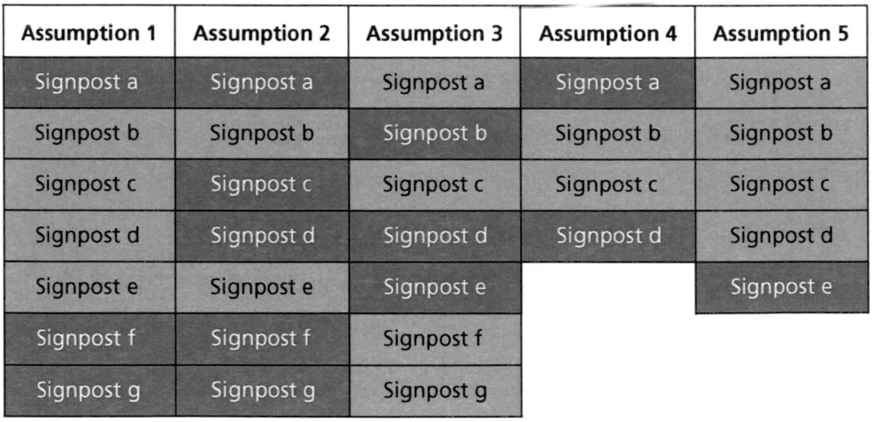

Assumption 1	Assumption 2	Assumption 3	Assumption 4	Assumption 5
Signpost a	Signpost a	Signpost a	Signpost a	Signpost a
Signpost b	Signpost b	Signpost b	Signpost b	Signpost b
Signpost c	Signpost c	Signpost c	Signpost c	Signpost c
Signpost d	Signpost d	Signpost d	Signpost d	Signpost d
Signpost e	Signpost e	Signpost e		Signpost e
Signpost f	Signpost f	Signpost f		
Signpost g	Signpost g	Signpost g		

RAND *MG532-A.2*

---

[3] However, the signpost assessment was meant to be reasonably credible. As noted in Chapter Two, the acquisition budget priorities the Army has established and those resulting from this research and AIM development are in general agreement.

indicators of emergent signposts. This tendency was resolved through an iterative process of assessment. The iteration allowed us to compare the assessments of signposts against each other, make the assessments more consistent, and correct for biases.

For assumption 1 concerning U.S. operational and technical dominance, signposts b through e are in sight. Signpost b appears when potential regional powers are demonstrating growing military capabilities and an inclination to increase their regional influence. India, for example, is in the process of acquiring a former Soviet aircraft carrier that would eventually allow New Delhi to project power well off its shores into the Indian Ocean. China's missile program, when it matures, can likewise be understood as ultimately giving Beijing the ability to dominate the South China Sea and Taiwan. China and India rank, respectively, the number one and number two arms importers for the period 2000–2004.[4] Iran's behavior, pursuing an indigenous missile and nuclear weapon program, might be viewed as consistent with the intent to dominate its neighbors and the Persian Gulf, although self-preservation might also motivate these actions. In addition, Tehran has imported significant numbers of anti-ship and anti-tank missiles, the former giving it the potential to threaten the Strait of Hormuz, the latter making Iran a potentially difficult land combat adversary.[5]

Signpost c, nuclear weapons and delivery systems appear in formerly non-nuclear states, emerged from Iranian, Pakistani, and Indian actions. Iran remains committed to its nuclear weapons program as does North Korea. France, Ukraine, Russia, and Pakistan have helped by providing "nuclear-capable" weapons.[6] Pyongyang, for its part, continues to defy the United States and the rest of the world as it progresses toward an operational nuclear force of its own.

Signpost d appears when key defense industries in the United States fail, stagnate, or consolidate. Whether or not this signpost is

---

[4] According to the Stockholm International Peace Research Institute (2005a).

[5] Stockholm International Peace Research Institute (2005b).

[6] See U.S. Naval Institute (2005). The term "nuclear capable" as used in this article means that the weapons described are suitable for carrying nuclear payloads, but the article does not address the question of whether they have such payloads.

emerging is a matter of some debate. Suzanne Patrick, until recently the Deputy Under Secretary of Defense for Industrial Policy, has argued that the industrial base is essentially healthy but would be better off if the Department of Defense made key investments to nurture critical capabilities. She even proposed a $100 million Defense Industrial Base Implementation Fund.[7] Her call for such a fund suggests that, at present, there are pressures on the industrial base; signpost d is emerging.

Signpost e deals with critical technology production moving overseas. According to the *Defense Industrial Base Capabilities Study* (U.S. Department of Defense, 2004c), some sectors of the industrial base in the United States are healthier than others.[8] Those areas with little redundancy in the number of U.S. suppliers are on a watch list. The watch list itself suggests that signpost e looms ahead.

Two signposts associated with the second, key, load-bearing assumption underpinning current acquisition investment strategy seem to be appearing. Recall that the second assumption is that the U.S. Army will require and maintain a capacity for rapid global deployment and self-sustainment in austere theaters. Signpost b, that HNS becomes questionable or contingent on the nature and location of U.S. operations, is clearly upon us. As evidence of this, consider Turkey's refusal to allow U.S. forces to transit its territory for a northern attack into Saddam Hussein's Iraq in March 2003. Germany and France both adamantly opposed the Iraqi campaign and undoubtedly would have refused U.S. requests for HNS to support operations against Iraq.

Signpost e, that technically feasible solutions for rapid deployment of U.S. Army assets are unavailable or unaffordable, seems clearly to be approaching. The time-distance problems inherent in global-range power projection dictate harsh terms. Only a significantly lighter, more compact force that can be lifted with the current airlift fleet—which is technically unavailable until FCS deploys—or a much larger, faster fleet of ROROs—which is probably unaffordable—can manage the

---

[7]   See "What Is the Real Health of the Defense Industrial Base?" (2005).

[8]   U.S. Department of Defense (2004c), especially the "watch list" of endangered capabilities (p. 47).

rapid deployment and self-sustainment challenges facing the Army. Signpost e confronts us head-on.

A majority of signposts for developments threatening the third assumption seem to be appearing. The third assumption expects that the Army will be increasingly adroit at managing complexity. Signpost a, continued and very heavy OPTEMPO severely constrains training opportunities for most Army units, is certainly with us. Current symptoms include the deployment of elements of the 2d Armored Cavalry Regiment (ACR), normally the opposing force at the National Training Center at Fort Irwin, for combat; disruptions to the professional military education schedule for both officers and enlisted soldiers; and, finally, the compression of the time frame from unit reset through training and redeployment into a theater of operations.[9]

Signpost c, concerning the Army's ability to recruit the right kinds of people, is the topic of recent press reports. According to these reports, the Army was 16 percent off its recruiting objective of 80,000 for FY 2005 as of the first of May that year.[10] If signpost c is not fully observable today, it surely looms on the horizon as an unwelcome but unstoppable development.

Signpost f, that technology readiness levels of critical technologies mature more slowly than expected, also seems to be emergent. The FCS—the heart of today's acquisition strategy—faces "unprecedented technical challenges" because of immature TRL for key technologies and because of the enormous challenge posed by systems integration across such a vast program.[11] Among the subsystems facing difficulties is the Joint Tactical Radio System (JTRS). This system lies at the very core of FCS, providing the connectivity to the Warfighter Information Network that ultimately links the system of systems together. Yet JTRS has been plagued with a host of problems including immature

---

[9] The Congressional Budget Office (2004) estimates the cost of increased OPTEMPO for the Army in terms of equipment depreciation at $4.68 billion.

[10] See The Peninsula (2005).

[11] See U.S. General Accounting Office (2004a).

technologies, underfunding by the services, and difficulties reaching consensus on performance criteria, among other things.[12]

Signpost g, that systems integration proves difficult, is closely linked to the previous signpost and also seems to be present today. FCS integration has proven much more difficult than anticipated and costs associated with systems integration have grown by almost 25 percent. In part because of systems integration considerations, the FCS program has been significantly restructured.[13]

Signpost h, that the civilian workforce continues aging, is also present in today's circumstances. Forty percent of current federal civil service employees will be eligible to retire within the next four years. Among high-level managers, the retirement eligibility sometimes crests 50 percent, depending on career fields.[14] Thus, according to the signposts emerging, the key, load-bearing assumption about the Army's ability to manage complexity is becoming vulnerable.

The fourth assumption underpinning Army acquisition investment strategy is that Army budgets will sustain today's operational and technical dominance. Three of the five signposts of vulnerability appear to be emerging. Signpost b, that OPTEMPO consumes a large percentage of the defense budget over time, describes today's circumstances. The U.S. General Accounting Office estimated that during FY 2003 alone, the global war on terrorism consumed $61 billion.[15] Further evidence that OPTEMPO is consuming ever-larger budget shares is the $82.04 billion emergency supplemental bill to fund the global war on terrorism and tsunami relief that passed into law in May 2005.[16]

Evidence in support of the notion that signpost c is emerging is less conclusive, but may become more so after the *Quadrennial Defense*

---

[12] U.S. General Accounting Office (2003).

[13] See a history of the program by GlobalSecurity.org (n.d.).

[14] Nelson (2004, pp. 202–215).

[15] See U.S. General Accounting Office (2004b).

[16] Details available at http://rpc.senate.gov/_files/May0905ConfRepHRes1268JG0.pdf

Review. The signpost that DoD-level decisions concerning missions reallocate resources in favor of the other services seems to be looming. As the Navy and Air Force continue to develop their capacities for land attack, these new capabilities encroach on Army roles and missions. If the other services successfully press their cases for rapid response and "over the horizon" presence, they may succeed in capturing resources that otherwise would have funded Army initiatives. Whether this signpost is fully present today or not, given the potential consequences for the Army, it bears careful scrutiny and management.

Signpost e, significantly increasing personnel costs, is certainly present today. Enlistment bonuses, paid in return for a two-year term of service, can reach $20,000, depending on career field. Monthly pay, not including housing allowances, rations, hazard, or hostile fire pay is $1,235.10 for a private with more than four months of service.[17] Add the enlistment bonus to the basic pay, various allowances, and health care benefits, and the entry-level Army salary package compares very favorably to other entry-level benefits packages: $26,393 for a claims clerk, or $26,182 for a receptionist.[18] Noncommissioned officers and officers with more extensive time-in-service earn proportionately more. Reenlistment bonuses have also been growing as the Army has to offer larger incentives to retain seasoned personnel during wartime. Reenlistment bonuses can, depending upon the military occupational specialty and member's years of service, approach $60,000.[19] The key, load-bearing assumption about adequate budgets to sustain operational and technical dominance is becoming vulnerable.

The fifth core assumption, that the Army can rely on the capabilities of its sister services and the Reserve Component, is also becoming vulnerable with four of its five signposts present today. Signpost a, that OPTEMPO challenges the ability of the reserves to train, is a matter of fact. The OSD 2005 *Reserve Component Employment Study* noted the training shortcomings and made a number of recommendations to address them. The report also noted the tension between homeland-

---

[17] Based on pay scales given in Military.com (n.d.b).

[18] Civilian salaries were found on Salary Wizard.

[19] See Military.com (n.d.a).

security-related missions and the Guard and Reserve roles in major theater wars and smaller contingencies abroad.[20]

Signpost b has also arrived. Evidence of this indicator, that multiple contingencies divert attention of Reserve Component units and the sister services away from Army priorities, lies in the multiple deployments that the United States supports around the world, the elevated OPTEMPO endured by all of the services and components that results from these contingencies, the difficulties the services face in retaining experienced personnel, and the increased incentives the services offer in an attempt to induce people to reenlist. Beset as they are with their own compelling issues and challenges, it is not surprising that the other services and Reserve Components pay less attention to Army priorities than they otherwise would.[21]

Signpost c, that sister services allocate their resources in ways that do not support Army priorities, also may be emerging. As the Army sheds its organic fire support as part of its transformation, it becomes more dependent on the other services to provide supporting fires. Thus far, however, there are no indicators that either the Air Force or Navy is modifying its force structure to produce more fire support for the Army. For example, neither service has added additional attack aircraft. Instead, the Air Force continues to pursue the F/A-22, which will have a limited air-to-ground capability.[22] The Air Force gunship fleet includes only 20 AC-130U/H aircraft, most of which support special operations forces.

The final signpost threatening the Army's assumption about the reliability of the other services and the Reserve Component is that neither is able to recruit and retain sufficient numbers of the right kind of people. This, too, seems to be increasingly the case. Although both

---

[20] U.S. Department of Defense (2005, p. 14).

[21] Even in 2000, the pressures were building from OPTEMPO. The Army was hoping to have the Reserve Component pick up every third or fourth rotation among the deployments under way at the time. At the time, Reserve Component enlisted attrition across the services was on the order of 30 percent and the Air Force Reserve was concerned about increased losses among mid-career personnel. See U.S. Congress (2000), especially the panel three testimony.

[22] According to the U.S. Naval Institute (n.d.).

appear to have made their aggregate goals for recruiting and retention, disaggregating the data reveals significant shortfalls in some military occupational specialties (MOSs). The U.S. General Accounting Office identified 16 MOSs in the Army Reserve that are consistently difficult to recruit and retain, 15 MOSs in the active Air Force, ten in the Air Force Reserve, and ten in the Marine Corps Reserve. Moreover, not all services met their goals. The Army Reserve made only 87 percent of its recruiting goal for FY 2004. The Air Force Reserve made 94 percent of its recruiting goal.[23]

It is clear from the analysis above that signposts suggesting vulnerability are present for each of the current, key, load-bearing assumptions. It is prudent, therefore, to consider alternative investment strategies and adjustments that might place the acquisition community on sounder footing for the future. As a part of this consideration, it is important to consider not only those areas where hedging and shaping are already necessary, but also future circumstances that might produce signposts of vulnerability that would have to be addressed. Because the time frame for major Army acquisition decisions can be lengthy from decision to implementation, and because signposts can appear with little or no warning, we determined that Assumption-Based Planning would benefit from an adaptation to increase its utility for the acquisition community. This adaptation involves considering alternative futures and developing an acquisition strategy that performs well in all of them. Appendix B examines the effect of alternative futures on the key, load-bearing assumptions.

---

[23] U.S. General Accounting Office (2005b, pp. 13–16).

Alternative Sets of Circumstances

This appendix describes the key features of the alternative sets of circumstances used to populate the Strategic Planning Guidance's construct of the global security environment in terms of conventional, irregular, disruptive, and catastrophic threats. These sets of circumstances are best thought of as being something like weather systems that can intrude into the present; some bring more hurricanes, others bring temperature extremes, and some bring drought. Their arrival, however, typically causes us to change our plans. And that is what we intend to describe here: new circumstances that, as they become more threatening, cause us to change our acquisition investment strategy—sometimes at the margins, other times more profoundly.

We offer six sets of circumstances, described in more detail below. Each description provides a general profile and key details inherent in the circumstances, and our expectations of Army missions and involvement in response to the circumstances.

Army Missions and Involvement in Regional Anarchy

Regional anarchy describes circumstances in which a confluence of demographic pressures, falling raw materials prices, antiquated infrastructure, corruption, and competing political or tribal loyalties creates a set of failed states and ungoverned zones in the developing world. Most of these anarchic regions are in Sub-Saharan Africa, but some could emerge in such areas as Central Asia, western Pakistan, and

Southeast Asia, where tribal and ethnic loyalties often supersede allegiance to the state.

The growth of anarchic zones generates a vacuum that is typically filled by charismatic warlords, such as Charles Taylor, the Liberian. These warlords often have no grand ideological agenda; their focus is on self-enrichment only. They normally rely on criminal activities, such as the smuggling of drugs and weapons, extortion of civilians, and kidnapping for ransom to fund an exorbitant life style and private armies that keep them in power. These armies are usually no more than ragtag militias of very young men who have joined simply to get a stable source of food, shelter, and wages. Their level of personal loyalty to the warlord is usually low.[1] The tactical and operational sophistication of the warlord armies is normally quite limited. Most of these private armies will be armed with only basic personal weapons, automatic weapons, and rocket propelled grenades. Some may be able to manufacture improvised explosive devices, and an occasional armored car and shoulder-launched antiaircraft missile might be found as well.

Even though the warlord armies generally have little real military capability, they are often strong enough to co-opt certain weak regimes and assume de facto control of many failed states. States such as Guinea, Liberia, Sierra Leone, and the Democratic Republic of Congo could be vulnerable to this kind of takeover. These states could then serve as launching pads for criminal activity or as bases for terrorist groups.

As anarchic conditions unfold, dangerous consequences often follow. For example, as the public health infrastructure collapses in an anarchic zone, infectious disease pandemics become more likely. Additionally, when there is no effective government control of an area, the potential for ethnic and tribal conflict to spin out of control looms greater. It is important to note that anarchy can also be "contagious." Viable states in the affected regions (so-called firewall states) are often placed under tremendous pressure. The increase in criminal activity in the neighboring anarchic region raises the probability that increased

---

[1]   There are exceptions to this rule, however. In such areas as western Pakistan, where tribal loyalties are very strong, these private armies may be held together by powerful social bonds that can withstand decreases in pay and food allotments.

corruption will occur. Even more likely is that the criminal activity itself will spill over the borders. Refugees from the anarchy will require resources (food, water, medicine, security, shelter) from the host nation, which is often poor itself. In these circumstances, neighboring states are likely to appeal to the United States, other Western states, and the United Nations for help in dealing with the emergence of anarchy.

In general, Army involvement is not deep, although just what role the Army fills depends significantly on the states in jeopardy. Typically, these circumstances find the Army performing non-combatant evacuation operations for U.S. embassy personnel and personnel representing friendly countries, providing humanitarian relief, conducting stability and support operations, and perhaps providing border security. In some instances, Army involvement may include refugee management and resettlement.

Army Missions and Involvement in Regional Insurgency

The circumstances here involve a variety of regional insurgent groups that threaten governments and ideals that are friendly to the United States. Many, but not all, of these guerrilla groups are Islamist, but rather than being part of a broader, global Islamist movement, their objectives tend to be parochial, emphasizing independence or autonomy. Some are devoted to old-fashioned Marxist and Maoist ideologies; others are driven by ethnicity and identity politics. Occasionally, these diverse groups might share tactical lessons learned and provide weaponry to each other, but there is no formal strategic cooperation between them. They are pursuing largely independent agendas.

The major insurgent groups in this future receive very limited state sponsorship, instead depending on financial support from individual donors, diasporas, religious foundations, charities, kidnapping for ransom, smuggling, and fraud.

The regional insurgent groups threatening American interests conduct intense guerrilla campaigns against the government forces of Colombia, Central Asia, Iraq, Israel, Egypt, Russia, Sri Lanka, and Nepal. They all conduct operations in both urban and rural areas. None has a real state sanctuary. Periodically, one of these groups mounts a spectacular terrorist strike in a major Western city to draw attention to its cause.

Regional insurgencies will engage the Army in a manner similar to that under way today, although perhaps more intensively and involving the Army in a greater number of foreign countries. Because these conditions include an increased threat to homeland security, the Army might find itself more deeply involved in the direct defense and security of U.S. territory.

Abroad, the Army would play an expanded role in foreign military and police training, foreign internal defense operations, perhaps some large-scale counterterrorism operations either unilaterally or in conjunction with a coalition or the host nation, and additional special operations including direct action against high-value targets.

Army Missions and Involvement in Broad Islamist Insurgency and Terrorism

Broad Islamic insurgency and terrorism involves a global jihadist campaign against the West in general and Israel and the United States in particular. It has the potential to become more organized, cohesive, and threatening over time. The long-term goal of the jihadist terror movement is a caliphate, initially in areas that are predominantly Islamic but eventually on a more global scale. Shorter-term goals include the toppling of corrupt or West-leaning governments in the Islamic regions of the world, the expulsion of Western interests from Islamic states, the destruction of Israel, and the establishment of fundamentalist Salafist governments in one or more states. The jihadist terror movement does not currently have any official state sponsors but there may be sympathizers within the governments of Islamic states. In addition, the jihadists enjoy some level of support among Sunni populations. This support is manifested in financial support, low-level assistance, and cover.

Lacking state sponsorship, the jihadists are unlikely to engage in conventional military operations. Instead, they will most likely use guerrilla and terror tactics to weaken governments and economies. The jihadist movement is organized into small cells that blend with target populations, although if a country, such as Afghanistan, can be controlled again by the jihadists or their sympathizers, larger groupings are possible. It is important to note that the cells that make up the jihadist movement are widely and globally dispersed.

In terms of technology, the jihadist movement does not attempt to make any real military technological breakthroughs other than the creation of the occasional dirty bomb for use in the West. Its focus is on organizational efficiency and covert communications techniques. At the tactical level, insurgent groups rely for most operations on a standard mix of automatic weapons, mortars, rocket-propelled grenades, low-altitude air defense weapons, improvised explosive devices, and crude rockets. Propaganda work that spreads the Wahhabi message on web sites and over satellite TV is also heavily emphasized.

The Army will encounter two categories of missions: (1) assisting friendly governments under pressure from Islamic insurgency and terrorism that threaten their ability to rule, and (2) direct, offensive action against insurgents and terrorists who threaten the United States, its allies, and its interests and regional objectives. Collectively, these activities will resemble the ongoing global war on terrorism, although they may expand into additional theaters of operation and in the size of Army forces deployed against the enemy. Missions in support of beleaguered friendly governments will include military training and assistance, foreign internal defense, and combined operations with host nation forces. Other missions may include unilateral offensive campaigns to find, fix, and destroy enemy forces operating in territory over which the local regime can no longer exercise its authority.

Army Missions and Involvement with Emergent Nuclear Powers

Emergent nuclear powers consist of a set of mid-range powers that have developed and deployed nuclear weapons. Some, but not all, of these powers are hostile to the United States and its allies.

A number of factors could significantly complicate the security issues associated with dealing with emergent nuclear powers. Some may enjoy cordial, or even militarily significant, relationships with world powers such as Russia and the People's Republic of China. Others may have failing governments or governments under significant internal threat. Finally, there are likely to be instances where two emergent nuclear powers are hostile to one another.

Three principal emergent nuclear power circumstances are particularly dangerous for the United States. In the first, an emergent nuclear power that is hostile to the United States uses its nuclear capability as a shield behind which it can safely engage in actions, overt and covert, in an effort to dominate a region or destroy an ideological, religious, or ethnic competitor. In the second circumstance, an emergent nuclear power is extremely hostile to, or nihilistic or fanatical in its abhorrence of, the United States. In such a case, a direct nuclear strike is conceivable. Finally, the government of the emergent nuclear power could lose control of weapons or technology. Espionage, criminal activity, civil war, or governmental collapse are all possibilities that could cause the loss of control. In any of these cases, terrorists or criminals could end up with nuclear weapons.

The threat to the United States posed by the emergent nuclear powers is substantial. We anticipate, therefore, that the Army would play a significant role in managing that threat, especially in preventing the emergence of still more nuclear powers. Army missions could include major conventional operations to destroy nascent nuclear weapons research efforts, including operations to seize research facilities, sensitive technical equipment, and nuclear materials. They might also include direct action against high-value targets implicated in the sale or distribution of nuclear technology, designs, and materials.

The Army would also play a major role in post-strike assistance, refugee management, and humanitarian relief in circumstances where one or more of the emergent nuclear powers actually detonates a nuclear weapon. The Army must also be prepared to conduct operations in a nuclear environment, including communicating in the immediate aftermath of an electromagnetic pulse, maneuvering through potentially contaminated terrain, and managing radiation exposure of its troops.

Army Missions and Involvement in Cold Peace Competition

Cold peace competition presents circumstances in which competition over resources and broad ideological differences could lead to a geopolitical influence-building contest around the globe between the United States, China, and Russia. None of the three states involved in this contest wants armed conflict to erupt; however, the intensity of the influence-building competition could lead to miscalculation or brinksmanship on the part of one or more actors, which could push the system down the slippery slope into outright war. The tension inherent in cold peace competition would also cause a return of strategic nuclear competition between the great powers.

The first major driver for cold peace competition is the competition for increasingly scarce petroleum and natural gas reserves. Chinese and Indian economic growth rates are dramatically driving up demand for energy. Since the United States will most likely not decrease its energy consumption, the result could be an ever-tightening world energy market with rising prices and growing fears of a supply shortage. As supplies tighten, competition for oil concessions, pipeline routes, and port access in oil-producing regions will also increase. To buttress economic influence in key oil-producing regions and to safeguard supplies, the major powers will draw on their security tools, such as arms sales, emplacement of military bases, mutual defense treaties, forward presence, employment of proxy forces, military aid, and combined exercises.

Russia is also affected by the resource competition because rising oil and gas prices are a boon to the Russian economy. In these circumstances, Russia, flush with foreign exchange reserves, could also begin an ambitious rearmament program and its own geopolitical influence-building campaign in its old sphere of influence.

The second major driver for cold peace competition is ideology. U.S. pressure around the world to increase democratization and ensure basic human rights is resented to some extent by the Chinese and Russian governments. They see this effort as a threat to their own systems and values and also as a grave danger to many of their most loyal allies.

Cold peace would find the Army heavily involved in some dimensions of the competition, and less so in others. The Army would participate extensively in brigade-level combined training exercises with countries Washington is seeking to influence. These will place a premium on mobile, expeditionary forces that have enough combat punch to defend and sustain themselves for one to two weeks in the event that localized combat breaks out. Air-deployable light infantry and medium-weight mechanized forces (such as the U.S. Stryker brigades) are invaluable. Additionally, because of its leading role, the Army is likely to increase its investment of money and resources in national missile defense.

Army Missions and Involvement with Rising Conventional Powers

Rising conventional powers would include two major powers defying the conventional wisdom of the U.S. national security community and undertaking ambitious efforts to build a full-spectrum air, ground, and naval conventional force with advanced command, control, communications, computers, intelligence, surveillance, and reconnaissance (C4ISR) that could challenge the United States directly in a conventional war in their home regions.[2] These actors might also have limited power-projection capabilities. They would consciously eschew the asymmetric strategy paradigm, instead preferring to develop themselves into traditional conventional military powers. Needless to say, the objective of this approach would be to establish an ironclad sphere of influence in the new power's home region.

---

[2]   We elected to use two rising powers to capture the old *Defense Planning Guidance* notion of facing two nearly simultaneous major theater wars.

What is most interesting about this set of conditions is that a wide range of states could conceivably embark on this path. Emerging great powers such as China and India, a regional rogue state such as Iran, or a traditional non-military great power such Japan could all choose this option. They all have the technological and financial wherewithal to take this path, albeit with significant stress to their economies. Some would view Russia as another candidate for this approach, especially if large revenues from energy exports boost the Russian economy over the long run. However, in our judgment, Russia is not a good candidate to become a rising conventional power because of its extremely long borders and stagnant to declining population.

For our purposes (i.e., to more completely populate the threat-likelihood space), we assume that one power emphasizes air and naval capabilities (China is probably the best candidate for this) and one pursues a fully balanced approach of modernizing air, ground, and naval forces equally (Iran is an excellent candidate for this balanced approach). It is possible that China and Iran might work together, at least covertly, to share tactics and technology in an effort to defeat their mutual American enemy. This would compound the problem for American defense planners immensely.

Combatant Commanders contemplating rising conventional powers within their areas of responsibility will in all likelihood count on significant Army forces in their contingency plans. To confront rising conventional powers, the Army would generate force packages specifically for regional contingencies, move forces forward into the theater of operations, and engage in multinational planning and exercises to organize regional states in resisting the growing power and influence of the would-be regional powers.

The resulting Army forces would not necessarily be "expeditionary" in our current understanding of that term, because they would not necessarily have to deploy from the United States. Given favorable strategic alignments, Army forces might operate from well-developed garrisons in or near the region with relatively short time-distance problems to overcome in the event of an emergency.

Moreover, the vastness of the potential theater of operations and the density of rival forces within them might overwhelm current expectations about battlespace awareness and the Army's ability to avoid enemy fires. A very different Army in terms of doctrine and equipment might be the result: something heavier than FCS with more emphasis on semi-autonomous operations.

Budget Categories

As the main text of this monograph indicated, we concluded that we needed a set of budget categories that would allow us to differentiate among investment types in more detail than is possible with the official Army budget categories (e.g., ammunition, armored vehicles, other vehicles, etc.) to see how funding might move in response to hedging and shaping decisions as alternative futures became more likely or more dangerous. We therefore constructed a set of budget categories similar in concept to the Army's old Battlefield Operating System categories.

We originally envisioned ten budget categories: close battle, mobility, direct fire, indirect fire, forward support, force protection, remote support, command and control, RSTA, and systems integration. As the research progressed, we added two more—one to account for funds we could not classify (these amounted to less than 10 percent of the Army's overall budget) and one for congressional additions not requested by the DoD or the Army.

Close Battle

This budget category includes accounts that support close battle: the fight inside 400 meters. It includes such equipment as body armor, night vision goggles, and other items that contribute to the close fight.

Mobility

This budget category includes accounts that produce mobility. If a piece of equipment's primary function is to move men or materiel (e.g., a five-ton truck, a UH-60 helicopter), it falls into this category. The fact that these platforms are typically armed for self-defense does not move them from this category into any other (direct fire, forward support).

Direct Fire

This budget category includes accounts for direct fire weapons: small arms, machine guns, and larger-caliber direct fire weapons including anti-tank guided missiles. The category also includes tanks and similar vehicles that serve as platforms for direct fire gun and missile systems.

Indirect Fire

This budget category includes accounts for indirect fire weapons: mortars of all calibers and artillery.

Forward Support

This budget category includes diverse accounts meant to provide combat and combat service support forward in the theater of operations. It includes engineer systems, logistics support, medical support, and generally any of the capabilities found in forward support battalions, DISCOMS, and COSCOMS.

Force Protection

This budget category includes accounts meant to provide force protection, ranging from air defense systems to immunizations to barrier materials. If the primary function of a system contributes directly to force protection, it is included in this budget category.

Remote Support

This budget category includes accounts that support Army forces from afar. It includes the TRADOC schools system, distance learning, and training support. It also includes the support provided by CONUS depots, laboratories, analytical centers, and all of the combat support and combat service support provided from somewhere outside the immediate theater of operations.

Command and Control

This budget category includes accounts that provide command and control capabilities. It includes communication systems, battle management systems, situational awareness systems, and data display systems that contribute to a commander's ability to understand the battlespace and maneuver forces over and through the battlespace for advantage over the enemy.

RSTA

This budget category includes accounts that provide reconnaissance, surveillance, or target acquisition capabilities. It includes radars, ground sensors, aerial sensors, and intelligence production, fusion, and transmission systems, e.g., ASIS, SOCRATES.

Systems Integration

This budget category includes accounts that contribute to systems integration.

Historical Approaches: The Interwar Era

The interwar period was one of great strategic uncertainty, rapid technological change, and small budgets for the U.S. Army. Today's Army is obviously very different from the interwar Army and America's role in the world has dramatically increased. Nevertheless, enough similarities exist between the two periods for the lessons of the interwar era to be germane to today's Army. The modern Army also faces a wide range of possible future threats and contingencies, confronts a time of rapid technological change (especially in the information sciences), and now faces the specter of stagnant or declining acquisition budgets for several years.

In the wake of the rapid post–World War I demobilization, there was intense debate in Congress over what form the postwar Army should take.[1] The Army leadership entered this debate with a basic belief that its service had performed well on the Western Front in 1918. Most senior officers felt that that the Western Front experience had validated Army assumptions and doctrine.[2] They felt that the experience of 1918 proved the primacy of infantry on the modern battlefield and the value of open, as opposed to static, warfare. Perhaps just as important, the Army leadership saw in the World War I experience a resounding validation of the citizen soldier concept. This was the idea that a fairly well trained militia force of part-time soldiers could perform superbly

---

[1] Odom (1999, pp. 14–16).

[2] Odom (1999, pp. 37–38).

against the best foreign armies when given some supplementary training before combat and led by a professional officer corps.

As Congress and the War Department considered the proper shape of the post–World War I Army, they also had to assess the potential threats the country would be facing in the next 10–15 years. Although the security environment looked fairly placid from the perspective of Washington in 1919, there were some concerns. In the near term, there was a clear requirement to defend the Mexican border, as that frontier was beset by banditry and spillover from the various factional conflicts that plagued post-revolutionary Mexico.[3] At the same time, the Army would need to supply some small units to police America's colonial possessions in the Far East and Latin America. The Philippines required a garrison of about 9,000 troops, whereas Tientsin, China, and Panama required smaller garrisons. Over the longer term, there was a fear, especially in the Navy, that Japan, fresh from acquiring some of Germany's imperial possessions in the Pacific, would soon be making a bid for hegemony in Asia.[4] This raised the specter that the Army might need to beef up its Philippine garrison at some point to repel a Japanese invasion. And, of course, the ultimate long-term worry was that another large war could erupt in Europe if the Versailles settlement and the League of Nations regime lacked international support. Another U.S. intervention in such a scenario could not be ruled out. Throughout the interwar period, tensions over the importance of these various threats would hamper Army modernization planning and investment strategy.

After several months of bitter acrimony between proponents of rival visions of the future Army, Congress finally passed the National Defense Act of 1920, which laid out the template for the interwar Army. This legislation fundamentally shaped the force from 1920 until 1935 but, unfortunately, there turned out to be a significant difference between what the legislation stated on paper and what presidential and congressional funding decisions would allow the Army to actually support in the field. In principle, the legislation created a fairly small Reg-

---

[3]   See U.S. War Department (1920, pp. 244–245).

[4]   See Miller (1991).

ular Army that would be kept at high readiness for sudden national emergencies. This highly ready force would be backed up, not by conscripts as some had wanted, but by two layers of citizen soldiers (the National Guard and the Organized Reserve).[5] The fondness for citizen soldiers that the Army brought out of World War I undoubtedly had an influence on this part of the National Defense Act. On paper, the act called for a Regular Army of 17,000 officers and 280,000 enlisted men, a National Guard of 435,000, and an Organized Reserve made up of men in Reserve Officers' Training Corps (ROTC) and military training camp programs around the country.[6]

However, administration and congressional preferences for keeping federal expenditures low soon gutted the force structure plan of 1920. By 1922, the Regular Army had shrunk to only 147,000 men in total.[7] The nine Regular Army divisions that were created by the 1920 legislation were kept, as the Army leadership absolutely refused to give up formal force structure; however, the low manning of the force meant that seven divisions were kept as skeletons only, and the remaining two were kept at full readiness to defend the Mexican border.[8] The decision was made that the border defense mission was important enough that the rest of the Army would be hollowed out if necessary to support it.

Doctrine development was slow and sporadic during the interwar period. Perhaps the best testament to this is the fact that the Army did not begin to change the square, four-brigade divisional structure of World War I vintage until the late 1930s, when the first major field exercises of the interwar period were showing that the square division was outmoded. The Army's treatment of armor during this era was conservative and ambivalent.

In terms of mobilization planning, the Army worked hard right from the end of World War I into the 1930s to do intensive and detailed

---

[5] Odom (1999, pp. 16–17).

[6] Crossland and Currie (1984, pp. 33–40).

[7] Odom (1999, p. 84).

[8] Weigley (1986, p. 259).

industrial and personnel mobilization planning that would serve as a template that could be used effectively during the next U.S. mass mobilization for war. The Army Industrial College was set up during this time, and Army logistics and ordnance leaders made it a point during the 1920s and 1930s to interact regularly with the chief executives of major industrial corporations to educate them about the needs of the Army in the event of another large war.[9] These actions would pay dividends for the service in World War II, as the industrial mobilization of the country went very smoothly and allowed U.S. forces to overwhelm their enemies with quantitative materiel superiority.

Overall, the Army's readiness was very low throughout the interwar period. The service would have required one to two years of intense preparation and refurbishment for it to be able to deal with even a moderately capable foe.

Interwar Army Threat Analysis

For most of the interwar period, the Army's intelligence branch (G-2) was unable to provide the service with a clear hierarchy of likely future threats. Army analyses of the period went back and forth between emphasizing the threat of light guerrilla-type forces in the Western Hemisphere and that of potential medium to heavyweight opponents in Europe and Asia. The result was a schizophrenic split in threat perception within the operational Army—one that hampered force and investment strategy development. One cannot fault Army intelligence solely for this shortcoming. Indeed, part of the blame lies with the various administrations of the time. Throughout the interwar period, the political leadership in both the White House and Department of State provided very little strategy guidance to the Army in terms of establishing priorities among threats and developing coordinated political-military strategies that could be used to counter specific threats.[10] As a result, Army intelligence was making its assessments without having

---

[9]   Weigley (1986, pp. 267–269).

[10]   See Boll (1988).

any real benchmarks from the national strategic leadership; in such a vacuum it is understandable that G-2 would have difficulties in doing clear, organized analysis.

Army threat analysis finally sharpened in focus in the mid-1930s, during the tenure of Chief of Staff Malin Craig, when German rearmament, the remilitarization of the Rhineland, and Italy's invasion of Ethiopia all combined to convince the Army leadership that the Fascist powers of Europe posed a clear and present danger to American security.

In the 1920s, Army G-2 reports devoted much time to analyses of the political turbulence in Mexico. The security of America's southwest border was a real concern at the time, and G-2 studied the intricacies of Mexican domestic affairs in great detail. The unsettled security situation in post-revolutionary Mexico presented four risks from the standpoint of Army analysts. First, large criminal gangs were operating with relative impunity in northern Mexico and these gangs would periodically cross the frontier (especially in Arizona and New Mexico) and threaten American border towns.[11]

Second, 1920s Mexico was a place of great political turmoil, much of which included Mexican Army participation. Senior Mexican generals frequently meddled in the affairs of the civilian government and these incursions into politics often climaxed with a coup attempt by a certain army unit.[12] The intra-Army fighting that followed often spread into areas along the U.S. border and the risk of conflict spillover was always present. Indeed, in the 1919–1920 time frame there are accounts in official Army congressional testimony of instances in which Mexican military aircraft pursuing retreating rebel army columns accidentally bombed U.S. territory along the border as a result of navigation errors. The frightening prospect of pitched battles between Mexican Army factions on U.S. soil warranted that G-2 keep close tabs on internal Mexican politics.

Third, although it may sound fanciful today, in the mid- to late 1920s, there was much trepidation within the Army intelligence com-

---

[11] See U.S. War Department (1920, pp. 244–245).

[12] One abortive coup attempt is described in HQ Eighth Corps Area (1927a, pp. 1–4).

munity over the possibility of hostile foreign powers gaining influence in Mexico and using that nation as a beachhead from which to threaten the United States. These scenarios do not appear to have been well developed at the time but did worry G-2 officers working along the Mexican border. One Army G-2 assessment written at Fort Sam Houston, Texas, in 1927 specifically mentions a threat scenario in which Mexico would conclude a secret agreement with Japan in which Japanese immigrants would be allowed to settle en masse in southern Mexico and Japan would be granted military bases in that same region. In exchange, Mexico would receive large amounts of Japanese investment to improve its economic situation.[13] At the time, it should be noted, Japan was having a problem with surplus population and there was a flow of Japanese immigrants into Argentina and Brazil. Also, the Japanese Navy of the 1920s, despite the restrictions imposed by the Washington Naval Conference, was busy fortifying bases in the Mariana Islands and aggressively looking for access options in other parts of the Pacific. Thus, G-2's worries about this scenario were not completely unfounded. However, there is no historical evidence that Tokyo and Mexico City ever contemplated such a deal at the time.

Fourth and finally, Army intelligence analysts believed that there was some moderate chance that Mexico might make a push to expand its influence in Central America at the expense of the United States. G-2 reports of the late 1920s highlight, for example, the outpouring of popular support in Mexico for the cause of the anti-U.S. Sandinista rebels in Nicaragua and the efforts of the Mexican government to use economic aid to build up its influence in Guatemala.[14]

Overall, all of these concerns notwithstanding, the G-2 reports reviewed for this research generally refer to the state of official U.S.-Mexican relations as positive during the 1920s and do not reveal much concern that Mexico's regime might turn decisively against the United States and become a declared enemy. Likewise, the Mexican Army was seen as having only a very limited conventional warfare capability. However, according to G-2, if worst came to worst in U.S. relations

---

[13] HQ Eighth Corps Area (1927b, pp. 1–3).

[14] See U.S. Army (1927a).

with Mexico, the Mexican Army did have some potent guerrilla warfare capabilities that could prove to be a headache for the U.S. divisions posted along the southwestern border.

Through most of the 1920s, G-2 assessments of the situation in Europe seemed to view the geopolitical environment there as stable but containing the seeds of future tensions if the Continent's diplomats failed to defuse the various border disputes and rival territorial claims that had emerged in the wake of the Versailles settlement. Germany was painted as being largely a responsible member of the European security community at the time, with no near-term ambitions to overturn the status quo. German Foreign Minister Gustav Stresemann was singled out as being a responsible and effective diplomat who was helping to keep Europe at peace through his ongoing dialogues with France and Poland.[15] G-2 assessments indicated great confidence in the ability of France and Britain to deter or defeat the Germans militarily, in the event of a sudden outburst of German nationalism and irredentism. G-2 reports on the state of the British and French Empires at the time generally viewed these two countries as having good future economic prospects and also as possessing a great abundance of resources in their colonies—resources that could not be matched by a Germany still struggling to recover from World War I.[16] German-Polish tensions over the location of the border drawn at Versailles were noted as a problem by G-2 and the risk of conflict escalation over that issue was recognized, but the consensus of the analysts seems to have been that Germany would, if necessary, simply not be willing to risk a small war over the question of its eastern border.

As the 1920s came to a close, it appears that G-2 perceived the USSR to be the most threatening European power. Army intelligence estimates argued that the idealist internationalism of the early Bolsheviks was being replaced by a more aggressive geopolitical stance on the part of the post-Lenin regime in Moscow.[17] Moscow was seen as acting in terms of shrewd geopolitical self-interest and G-2 argued that

---

[15] See U.S. Army (1929, pp. 12320–12323).

[16] See U.S. Army (1928, pp. 11928–11931).

[17] See U.S. Army (1927b).

Soviet goals were extremely expansive throughout all of Eurasia, to include the dismantling of the British imperial presence in India and the coercion of Turkey into becoming a satellite state. The language used to describe Soviet strategy and intentions was much darker and more foreboding than that used with respect to any other European or Asian power of the time.

In the early 1930s, the amount of time devoted in G-2 reports to Mexico declined as the Mexican political situation stabilized. The environment along the southwest U.S. border improved as a consequence. In terms of the Western Hemisphere, assessments of the time focused on reports of Communist infiltration of Latin American militaries (especially Brazil) and monitoring the activities of the various European immigrant communities in Latin America, who were viewed as potential sources of foreign influence in the U.S. backyard.

Army intelligence still did not seem to be too perturbed by the situation in Germany in the early 1930s. In fact, one G-2 assessment from 1932 argued confidently that the rise of the Nazi Party in the recent German legislative election was not a cause for worry because if Hitler took power he would likely pursue a foreign policy oriented toward the status quo that would preserve stability in Europe.[18] However, Army analysts did perceive Italy to be a rising and dangerous military force that could inflame the security environment in southern Europe over time. Numerous reports were filed on Italy's massive military mobilization system, which drew almost all able-bodied males into some kind of active or reserve unit in either the military or paramilitary forces.[19] One report noted with admiration that Italy possessed the most efficient military mobilization system in the world. Italian military maneuvers were described in great detail, especially those parts that employed motorized units, in the worldwide intelligence estimates of the time.

The strategic balance in Asia was covered more extensively in the early 1930s than it had been in the previous decade. G-2 officers carefully monitored the Soviet-Japanese military competition in the Far

---

[18] See U.S. Army (1932).

[19] See U.S. Army (1934).

East in the wake of Japan's 1931 move into Manchuria and some estimates argued that the outcome of the Soviet-Japanese rivalry would do
much to determine the global balance of power for the rest of the 20th
century.[20] In the early part of the 1930s, G-2 estimates hinted at a belief
that the USSR would probably triumph over Japan because of the sheer
quantity of resources that the Soviets were able to deploy in Siberia and
their Far Eastern province. However, toward the middle of the 1930s,
as Japanese naval power grew in the Pacific and Japan seemed poised
to invade China proper, Army intelligence analysts became more pes-
simistic about Soviet prospects against the surging Japanese.

By the mid-1930s, G-2 clearly realized that the combination of
Nazi Germany and Fascist Italy was becoming the main threat to
global stability and American security. Reporting on the growth of
the German military and German intelligence operations in Austria
and Czechoslovakia became more and more frequent in the biweekly
intelligence assessments produced for the Army and national strategic
leaderships. Italy's military buildup and invasion of Ethiopia also drew
extensive attention. After 15 years of uncertain threat priorities, the
Army finally had a likely enemy against which it could develop a comprehensive investment strategy. In the end, it was only the dramatic
developments in the international system in the mid-1930s that allowed
the Army to produce a clear threat forecast. The service's internal intelligence assessment process of the time simply did not have analytical tools strong enough to handle the ambiguity and uncertainty that
characterized the threat environment of 1919–1935. Thus, by default,
Army threat perceptions of the time were driven by the personal judgments of senior officers.

Resulting Threat Perceptions in the Operational Army

The diffuse Army threat analysis of 1919–1935 created a conflicted
mindset within the operational Army over what kind of future conflicts the service should prepare for. The inability of G-2 to develop

---

[20] See U.S. Army (1935).

clear priorities among light Western Hemisphere and heavier European and Asian threats allowed a virtual schizophrenia to develop within the operational Army as different parts of the service emphasized different kinds of threats as the basis for planning and budgeting.

The secondary literature on the interwar Army's doctrine and force structure reveals that the senior leadership of the Army (i.e., the Chief of Staff's office) tended to see light threats in the Western Hemisphere as the most immediate concern. In contrast, the "analytical Army," which was made up of blue ribbon boards and commissions, the faculty of the service schools, and the General Staff War Plans Division, viewed another major conventional conflict in Europe against a first-class "heavy" opponent as the contingency to be planned against. This conflict was not resolved until Nazi Germany's emergence as a full-fledged threat to the European order in the mid-1930s removed all doubt about what was the principal threat to the United States.

At the senior leadership level, the light threat paradigm was set in place by 1920, when the new Chief of Staff, General John Pershing, declared that, "Our army is most likely to operate on the American Continent . . ." during a speech on future threats to the nation.[21] Pershing went on to advocate the formation of a light, mobile army that focused on the American Southwest. Chief of Staff Douglas MacArthur carried on this tradition when, in 1930, he pushed for the creation of an Immediate Readiness Force that would be tasked to respond to hot spots in the Western Hemisphere, such as the chaotic and anarchic Caribbean islands. For good measure, MacArthur ordered the Army General Staff to deemphasize planning for the mobilization of the Organized Reserves—an option that would need to be used only in the event of a large conventional war in Europe or Asia. Finally, MacArthur cemented his preference for lighter contingencies by issuing a decree that all U.S. Army tank designs must be less than 15 tons and that the lightest armor possible should be used in their construction.[22] He believed that the best defense for tanks was their speed and mobility, which reflects a cavalry mindset with regard to the employment of armor—one that was cer-

---

[21] Weigley (1986, p. 257).

[22] Weigley (1986, p. 259).

tainly not suited for contingencies in which the United States would face a large European army.

While various Army Chiefs of Staff were advocating a focus on light threats, the analytical branches of the Army were going in a very different direction. They were making recommendations and taking actions that indicated a strong belief that the Army had to prepare itself for major conventional warfare, in which firepower, force size, and efficient mobilization would be the keys to success. This pattern began right after World War I, when the American Expeditionary Force (AEF) Superior Board on Organization and Tactics argued that the Army ought to keep the heavy, square divisional structure because of the massed firepower the four brigade construct could deliver.[23] Mobilization planning by the Army General Staff during the 1920s was extensive and detailed, as the historical record tells us that six separate national mobilization plans were produced between 1923 and 1936.[24] These plans included provisions for both industrial and personnel mobilizations on a massive scale. They were enabled by the creation, in 1921, by the Harbord Board of a new office in the War Department called the Assistant Secretary of War for Mobilization. This office served as a critical link between the General Staff and the civilian leadership of the War Department on mobilization issues, a link which had not existed during the botched mobilization effort of 1917–1918.[25] Finally, during the interwar period the Army War College maintained a curriculum that tasked students to play war games and engage in conflict simulations that were overwhelmingly Euro-centric in character.[26] Little time was devoted to studying Western Hemisphere threats. This fact was of more than academic value to the Army, because in the 1920s and 1930s, the Army War College had a more official role in war planning than it does today. The students there were seen as a de facto extension of the General Staff War Plans Division, which used the results of War

---

[23] Odom (1999, p. 19).

[24] Weigley (1986, p. 268).

[25] Weigley (1986, pp. 267–269).

[26] See Gole (1985, pp. 52–64).

College games and exercises in its own planning. Furthermore, many War College students went directly to the General Staff War Plans Division on graduation, which was the part of the staff that assembled national mobilization plans.

Overall, then, we see that the Army had a conflicted mindset over what kinds of threats it would face in the future. The threat analysis of the time was too diffuse to break the deadlock. Until Germany's arms buildup accelerated in the mid-1930s, neither side in the future threats debate was able to prevail and the result was an Army that remained oriented toward large wars almost by default but assumed bits and pieces of the light war paradigm, creating contradictions and hampering force development and investment strategy. There are several examples of this. Perhaps the most notable is the gridlock over tank development that persisted into the 1930s. The Army's confusion over whether to pursue heavy or light models further complicated a tank development process that was already dysfunctional as a result of being split between the Bureaus of Infantry and Ordnance. This confusion stymied the progress of all the new models that were proposed, leaving the Army with no real tank procurement until the mid-1930s. Another contradiction that developed was the long-term juxtaposition of a Regular Army force structure that was designed to be a highly ready, professional, small war force onto an Organized Reserve structure that was built into a construct appropriate for mass mobilizations and large wars. The inability of the Army to decide whether it wanted to focus its resources on either a small Regular Army or a large Reserve caused the whole force to be crippled by low readiness from 1920 to 1935–1936. Even within the Regular Army itself there were contradictions. The Regular force's notional mission of being a small, ready, rapid reaction force was belied by the fact that it maintained a very high officer-enlisted ratio throughout the interwar period, making it unbalanced and not suitable for rapid deployment. The ratio that was held in this period was indeed one appropriate to a cadre force being groomed to support a mass mobilization. Finally, the confusion over future threats led to a stagnation in Army force organization, as the square division construct remained in the Army because of sheer inertia until World War II. There simply was not enough consensus on the nature of future threats and contingencies to justify any experimentation with division-level force organization.

The Army's Interwar Investment Strategy

In view of that state of the interwar Army and its threat perceptions, we now assess the investment strategy approach that the service pursued during the interwar period. How did the Army's leaders attempt to shape investment strategy to cope with the mixture of geopolitical uncertainty, rapid technological development, and low funding levels that confronted them?

The Army did not have a single investment strategy during the interwar years, but instead its strategy shifted in response to changes in the external environment. One can discern three distinct phases in interwar investment strategy. The first was the 1919–1929 period, the second, 1930–1935, and the third, 1936–1939. Each strategy was a response to the particular exigencies of a part of the interwar era.

Before discussing the three strategies in turn, it is important to understand one underlying feature of the Army in the interwar years. This is that the Army leadership consistently held that force size was the most important element to preserve, ahead of both readiness and modernization spending. The Army leadership fought to keep its nine-division skeleton structure from the end of World War I up until another war in Europe loomed in 1935–1936. This unshakeable priority, which was based on a belief that the service needed to hold to a certain force size floor to be viable in either a light or heavy contingency, meant that the acquisition account (which included R&D) always took the brunt of the congressional funding cuts that occurred constantly. To protect manpower and personnel spending, weapon modernization and research were frequently cut to the bone, resulting in a series of tiny acquisition budgets (see Table D.1). Thus, Army investment strategy was resource-constrained to the point that shaping actions were impossible for much of the interwar period. Only basic hedging

Table D.1
Interwar Army Acquisition Budgets

Year	$ Millions
1930	23.2
1931	31.8
1932	32.5
1933	18.8
1934	8.9
1935	21.3
1936	54.3
1937	63.2

actions could be undertaken. It should be noted here that this analysis examines only the Army investment strategy for ground forces. The Army Air Corps of the interwar period had a much more generous acquisition budget to work with because of popular enthusiasm for aviation in the interwar period and a belief that long-range air power could help the Navy defend America's coasts from hostile naval forces. The Air Corps is a separate case that is not treated here.

The early interwar period (1919–1929) saw the Army adopt an "experimentation and observation" investment strategy. As postwar budgets declined precipitously, Army leaders put faith in the huge stocks of World War I weapons that remained in the inventory. They believed that these would allow the Regular Army and Reserves to do enough basic training to keep up standards of readiness.[27] Another factor weighing on the minds of Army leaders was the rapid pace of technological change ongoing in the early 1920s, especially in the automotive industry. They feared that if the Army went ahead and undertook full-scale new weapons procurement projects, it would be wasting scarce money on systems that might well become outdated by the time the Army faced its next major conflict. The upshot of all of this

---

[27] Odom (1999, p. 96).

was an investment strategy that comprised two principal components: small pilot weapon programs and an intense monitoring of those manufacturing advances in private industry that had potential relevance to defense production. This was an intelligent and prudent strategy that, unfortunately, did not really bear fruit for the Army.

During the 1920s, the Army oversaw a multitude of small pilot weapon research programs across a variety of capability areas, including tanks, artillery of all calibers, antiaircraft artillery, mortars, and semiautomatic rifles.[28] As time went on this effort produced a number of prototypes in each area, most of which had very little in common. As an example, in the armor field, the Army's Bureau of Ordnance produced no less than 12 different tank prototypes during the decade. All of this small-scale engineering work gave the Army very little in the way of new capabilities or even new technological concepts. For a variety of reasons, almost none of the 1920s pilot programs went into mass production or even low-rate production.[29] In most cases, a lack of funding stopped work at the research phase, whereas in others, it was the existence of those large stocks of surplus weaponry from World War I. Tank development was stymied largely by organizational inefficiencies resulting from the split in development oversight between the Ordnance and Infantry Bureaus. The Infantry Bureau was assigned the task of setting up general specifications for each tank prototype and Ordnance had complete autonomy in translating the specifications into an actual combat vehicle.[30] In practice, this meant that the Ordnance Bureau took no input at all from the Infantry during the whole design and manufacturing process; it did not solicit Infantry inputs until the day the new prototype was actually delivered to the Infantry Branch Headquarters. Naturally, this was a flawed process that ended with the Infantry rejecting all the new prototypes of the 1920s.

At the end of the 1920s, the procession of failed and stalled prototypes left the Army with essentially the same capital stock it had possessed at the end of World War I. The turn of the decade saw the

---

28 Odom (1999, p. 211).

29 Weigley (1984, pp. 409–411).

30 See Johnson (1990, pp. 118–120).

Army still using the 1903 Springfield rifle as its main infantry weapon. The Stokes mortar was still in the inventory as well. The principal anti-tank weapon of the U.S. Army remained .50-caliber machine guns, despite the fact that the new tanks being fielded in some European armies made this weapon obsolete. The Artillery branch still relied on the French 75 mm gun as its bread and butter weapon and the tank inventory was unchanged from 1919, with the bulk of the roster being composed of 1,000 French Renault light tanks and British Mark VIII heavy tanks. Perhaps most damaging of all was the shortfall in motor-ized transport, where the Army leadership estimated that the service needed over 9,000 new trucks to be mission-capable.

The second part of the Army's investment strategy of the early interwar period was to observe new industrial processes with possible military utility in preparing for a future mass industrial mobilization. This effort seems to have encountered more success than the weapon pilot programs did, as it laid the foundation for the mobilization plans that proved so valuable during World War II. Congressional testimony of Army leaders in the 1920s mentions a number of industrial processes of interest.[31] New cold working techniques for building gun barrels, the introduction of reusable liners for field and antiaircraft guns, and the use of centrifugal casting instead of forging for gun barrels are all discussed with enthusiasm during congressional hearings on the annual state of the Army.[32] Perhaps as important as the monitoring of industrial processes themselves were the relationships fostered between the Army's ordnance and acquisition officer corps and the chief execu-tives of major manufacturing companies, as the Army conducted its active monitoring of innovation in American industry. This web of contacts simply had not existed during World War I and its absence crippled the American mobilization effort in that conflict. In the inter-war period, the Army did an excellent job of making sure that it cul-tivated the leaders of American industry, making them aware of the responsibilities they would incur should the nation enter into another large conventional war.

---

[31]   U.S. War Department (1929, pp. 57–65).

[32]   U.S. War Department (1929, pp. 57–65).

In the middle interwar period (1930–1935), the Army's investment strategy can be characterized as defensive and reactive. These were the darkest years of the Great Depression and the already small acquisition budgets of the 1920s shrank even further as the Hoover and Roosevelt administrations desperately tried to funnel government spending toward programs that would ease the economic crisis. The Army simply had no real means of embarking on a serious modernization effort in these years. The service did what one would expect, namely, it protected its foundational elements and starved everything else of funding. As one might guess, the Army focused on protecting the Regular Army's force size, officer-enlisted ratio, and the basic structure of the Organized Reserve. These were the bedrocks of the service in the view of its leaders and all resources were devoted to sustaining them until the economic climate improved. The pilot programs of the 1920s disappeared and only the Army Air Corps received significant procurement monies.

Finally, it is interesting to note what the Army's major acquisitions would have been in the middle interwar period had enough funds been available to actually make some significant purchases. We can get an idea of what the service's priorities were from a "wish list" that was produced by Chief of Staff MacArthur in 1930.[33] MacArthur said that, in an ideal world, the Army would request 18,000 new trucks to motorize the whole Army, an updated 75 mm artillery piece, new anti-aircraft artillery with advanced fire control, and a semiautomatic rifle.[34] The emphasis in the 1930 wish list was on incremental modernization, not on any attempt at radical innovation. The Army leadership did not request a 105 mm artillery piece or any advanced new tank designs, instead focusing on basic systems that represented small advances in capability over the previous generation.

In the late interwar period (1936–1939), the Army saw clear signposts that foretold of geopolitical problems ahead. The rapid German military buildup, the remilitarization of the Rhineland, and the Italian invasion of Ethiopia all pointed to an increased risk of war in Europe.

---

[33] Johnson (1990, pp. 250–264).

[34] Odom (1999, p. 108).

In response, Chief of Staff General Craig decided that the Army had to make hard decisions sooner rather than later to prepare for the possibility that the United States might be dragged into a major European war. Concerns over light threats were cast aside.

Craig decided that the Army's top priority would be to equip and man at least a few divisions to the point where they would be ready to deploy abroad with some basic capabilities. In 1936, at the beginning of this period, such a force just did not exist in the U.S. Army. To accomplish this goal, Craig was forced to drastically curtail the Army's R&D expenditures so as to concentrate money on the acquisition of immediately available weapon systems.[35] Craig was pursuing a "short-term readiness" investment strategy that emphasized buying sufficient quantities of what was available from American industry and Army arsenals, even if that meant fielding forces that were equipped with previous-generation weapons. This strategy is perfectly rational when seen in the context of the time; General Craig was reading the signposts of his era and calculated that he might not have time to wait for technologically advanced weapon systems to become available in bulk. Conflict might erupt before the latest systems were ready and the Army had to have some conventional combat capability if and when war started. The almost complete lack of modernization of the Army's capital stock in 1919–1935 period was forcing the service to choose the "least bad" from a series of unsatisfying investment strategy options.

The R&D cuts of 1936 had a damaging effect on the Army in the early years of World War II. This policy froze many weapon systems in the mid-1930s at a time when the Wehrmacht was fielding weapons that were state of the art. The effect of the 1936 decision was felt by the Army in the North African campaign of 1942–1943, as American units entered combat against the Wehrmacht with many antiquated weapons, such as the Stuart light tank, the 37 mm antitank gun, and old, lightly armored half-tracks.[36] Indeed, even the early models of the Sherman medium tank had deficiencies when matched up against the German Mark IV medium tank. As a result of these equipment short-

---

[35] Johnson (1990, p. 264).

[36] For a review of the North Africa campaign, see Atkinson (2002).

comings, U.S. forces in North Africa suffered heavy casualties in early engagements with the Germans such as at Happy Valley and Kasserine Pass. In the latter case, U.S. forces also suffered an ignominious battle-field defeat.

Lessons for Today's Army Acquisition Community

The leadership during the interwar era was hobbled by a lack of funding, and the acquisition community itself was poorly organized. That said, given the diverse visions of what the Army should become, the individual preferences of influential leaders such as Generals Pershing and MacArthur, and the divergent threat assessments of the time, it is doubtful that the acquisition system of the day would have produced an Army more suitable for the early engagements of World War II had the funding been there throughout the 1920s and 1930s. Divergent design points—light or heavy contingency, a campaign in the Americas or further abroad—and no systematic way to assess, appraise, and harmonize acquisition decisions, compounded by a profound lack of resources, resulted in an Army suboptimized for all contingencies.

Today's acquisition community need not suffer from the pathologies that affected interwar acquisition decisions. In addition to enjoying richer resources than their predecessors did, today's acquisition planners can make use of planning tools that will show them the consequences of their decisions before they make them and thus allow them to choose a course of action likely to be robust given today's uncertainties.

Bibliography

"Army Modernization Plan 2005: Call to Duty—The American Solider in Service to Our Nation for 230 Years." As of February 22, 2007:
http://www.army.mil/features/MODPlan/2005/MP2005CoverFront.jpg

Assistant Secretary of the Army (Acquisition, Logistics, and Technology), "Army Science and Technology Master Plan," Defense Technical Information Center (DTIC). As of February 22, 2007:
http://stinet.dtic.mil/oai/oai?&verb=getRecord&metadataPrefix=html&identifier=ADM001246

Atkinson, Rick, *An Army at Dawn: The War in North Africa, 1942–43*, New York, N.Y.: Henry Holt & Co., 2002.

Boll, Michael M., *National Security Planning: Roosevelt Through Reagan*, Lexington, Ky.: The University Press of Kentucky, 1988.

Congressional Budget Office, 2004. As of February 22, 2007:
http://www.cbo.gov/showdoc.cfm?index=6235&sequence=0

Crossland, Richard B., and James T. Currie, *Twice the Citizen: A History of the U.S. Army Reserve, 1908–1983*, Washington, D.C.: Office of the Chief, Army Reserve, 1984, pp. 33–40.

Davis, Paul K., Jonathan Kulick, and Michael Egner, *Implications of Modern Decision Science for Military Decision-Support Systems*, Santa Monica, Calif.: RAND Corporation, MG-360-AF, 2005.

Defense Acquisition Guidebook, DoD Directive 5000.1. As of December 19, 2005:
http://akss.dau.mil/dag/TOC_GuideBook.asp?sNode=R&Exp=Y

Defense Acquisition Guidebook, DoD Directive 5000.2. As of December 19, 2005:
http://akss.dau.mil/dag/TOC_GuideBook.asp?sNode=R&Exp=Y

Dewar, James A., *Assumption-Based Planning: A Tool for Reducing Avoidable Surprises*, Cambridge, U.K., and New York, N.Y.: Cambridge University Press, 2002.

DoD Instruction 5000.2. As of December 19, 2005:
http://akss.dau.mil/dag/TOC_GuideBook.asp?sNode=R&Exp=Y

Eaves, B. C., "Finite Solution of Pure Trade Markets with Cobb-Douglas Utilities," *Mathematical Programming Study 23*, 1985, pp. 226–239.

Enthoven, Alain C., *How Much Is Enough? Shaping the Defense Program 1961–1969*, Santa Monica, Calif.: RAND Corporation, 2005.

Gale, D., *The Theory of Linear Economic Models*, New York, N.Y.: McGraw-Hill Book Company, 1960.

GlobalSecurity.org, "Future Combat Systems (FCS)," n.d. As of February 22, 2007:
http://www.globalsecurity.org/military/systems/ground/fcs.htm

Gole, Henry G., "War Planning at the War College in the Mid-1930s," *Parameters*, Vol. XV, No. 1, Spring 1985.

Headquarters, Department of the Army, "Operations," FM 3-0, June 2001. As of February 22, 2007:
http://www.dtic.mil/doctrine/jel/service_pubs/fm3_0a.pdf

———, "Future Combat System (FCS) Operational Requirements Document (ORD)," April 14, 2003. As of February 22, 2007:
http://www.army.mil/fcs/factfiles/overview.html

Headquarters, Department of the Army, Office of the Deputy Chief of Staff G-3 (DAMO-SSW), Total Army Analysis documents. HQ Eighth Corps Area, "Summary of Intelligence, 9/28/27," Fort Sam Houston, Tex., 1927a.

———, "Summary of Intelligence, 11/23/27," Fort Sam Houston, Tex., 1927b.

Johnson, David, *Fast Tanks and Heavy Bombers*, dissertation, Duke University, Durham, N.C.: UMI Dissertation Services, 1990.

Military.com, "Enlistment and Reenlistment Bonuses," n.d.a. As of February 22, 2007:
http://www.military.com/Resources/ResourcesContent/0,13964,30965,00.html

———, "Military Pay Overview," n.d.b. As of February 22, 2007:
http://www.military.com/Resources/ResourcesContent/0,13964,49020,00.html

Miller, Edward S., *War Plan Orange: The U.S. Strategy to Defeat Japan: 1897–1945*, Anapolis, Md.: U.S. Naval Institute Press, 1991.

Nelson, Steve, "The State of the Federal Civil Service Today," *Review of Public Personnel Administration*, Vol. 24, No. 3, September 2004.

Odom, William O., *After the Trenches: The Transformation of Army Doctrine, 1918–1939*, College Station, Tex.: Texas A&M University Press, 1999, pp. 14–16.

Office of the Secretary of Defense, "Aldridge Study," Washington, D.C., 2005.

————, *The Future Year Defense Plan (FYDP)*, Washington, D.C., 2006.

The Peninsula, "Pentagon Delays Releasing Recruitment Data," June 2, 2005. As of February 22, 2007:
http://www.thepeninsulaqatar.com/Display_news.asp?section=World_News& subsection=Americas&month=June2005&file=World_News200506023729.xml

"Research, Development, and Acquisition Plan (RDAP)" in Chapter 11, "Materiel System Research, Development, and Acquisition Management," published by the U.S. Army War College (24th ed., 2003–2004, July 27, 2005). As of December 19, 2005:
http://carlisle-www.army.mil/usawc/dclm/linkedtextchapters.htm

Salary Wizard, homepage, September 15, 2005. As of October 7, 2005:
http://swz.salary.com/salarywizard/layoutscripts/swzl_compresult.asp? NarrowCode=OF01&NarrowDesc=Administrative%2C+Support%2C+and+ Clerical&JobTitle=Receptionist&JobCode=OF13000017&geo=U.S.% 20National%20Averages

Stockholm International Peace Research Institute (SIPRI), *2005 Yearbook*, 2005a. As of February 22, 2007:
http://www.sipri.org/contents/armstrad/at_data.html

————, special report on states of interest, 2005b. As of February 22, 2007:
http://www.sipri.org/contents/armstrad/atira_data.html

U.S. Army, G-2 Intelligence Summary, "Guatemalan-Mexican Relations," College Park, Md.: National Archives, September 2, 1927a.

————, "The International Position of the Soviet Union," College Park, Md.: National Archives, October 1, 1927b.

————, "France: 1927 in Retrospect," College Park, Md.: National Archives, January 6, 1928, pp. 11928–11931.

————, "German-Polish Boundary Question," College Park, Md.: National Archives, January 4, 1929, pp. 12320–12323.

————, "If Nazis Gain Control," College Park, Md.: National Archives, January 1, 1932.

————, "Italy's Military Preparation of the Citizen," College Park, Md.: National Archives, December 28, 1934.

————, "Soviet-Japanese Rivalry in the Far East," College Park, Md.: National Archives, January 10, 1935.

U.S. Army War College, "Planning Phase," Section 9-5, Chapter 9; "Army Planning, Programming, Budgeting, and Execution System," 24th ed., 2003–2004, July 27, 2005a. As of December 19, 2005:
http://carlisle-www.army.mil/usawc/dclm/linkedtextchapters.htm

————, "The MDEP: What It Is and How It's Used," Section 11-85, "Research, Development, and Acquisition Plan (RDAP)," 24th ed., 2003–2004, July 27, 2005b. As of December 19, 2005:
http://carlisle-www.army.mil/usawc/dclm/linkedtextchapters.htm

————, "Materiel System Research, Development, and Acquisition Management," Chapter 11, 24th ed., 2003–2004, July 27, 2005c. As of December 19, 2007:
http://carlisle-www.army.mil/usawc/dclm/linkedtextchapters.htm

————, "PPBES Deliberative Forums," Section VII, Chapter 9, in "How the Army Runs," 24th ed., 2003–2004, July 27, 2005. As of December 19, 2005:
http://carlisle-www.army.mil/usawc/dclm/linkedtextchapters.htm

U.S. Congress, *Sustaining the All-Volunteer Force and Reserve Component Issues,* Congressional Hearing Summary, March 8, 2000. As of February 22, 2007:
http://www.hqda.army.mil/ocll/Hearing_Summaries/03-08-2000_HASC_P_Sustaining_the_All_Volunteer_Force_and_Reserve_Component_Issues.htm

U.S. Department of Defense, "A Framework for Strategic Thinking," briefing for Senior Level Review Group, August 19, 2004a. As of December 19, 2005:
http://www.washingtonpost.com/wp-srv/nation/documents/defense_aug_19_2004.pdf

————, *Strategic Planning Guidance, Fiscal Years 2006–2011,* November 2004b. As of December 19, 2005:
http://www.dote.osd.mil/reports/TestinginaJointEnvironment-Public111204.pdf

————, "Defense Industrial Base Capabilities Study: Protection," December 2004c. As of December 19, 2005:
http://www.acq.osd.mil/ip/docs/dibcs_protection_12-28-04.pdf

————, *National Defense Strategy of the United States of America,* Washington, D.C., March 2005.

————, "Planning, Programming, Budgeting and Execution Process," Section 1.2, Chapter 1, "DoD Decision Support Systems," *Defense Acquisition Guidebook* (DAG), 2006. As of December 19, 2005:
http://akss.dau.mil/dag/GuideBook/PDFs/Chapter_1.pdf

————, "Reserve Component Employment Study 2005," Volume 1, "Study Report," Washington, D.C., 2005, p. 14. As of February 22, 2007:
http://www.fas.org/man/docs/rces2005_072299.htm

U.S. General Accounting Office, "Challenges and Risks Associated with the Joint Tactical Radio System Program," GAO-03-879R, August 11, 2003. As of December 19, 2005:
http://www.gao.gov/new.items/d03879r.pdf

———, Future Combat Systems Challenges and Prospects for Success, GAO-05-442T, March 15, 2005a. As of October 7, 2005:
http://www.gao.gov

———, *Military Operations: DoD's Fiscal Year 2003 Funding and Reported Obligations in Support of the Global War on Terrorism*, GAO-04-668, May 2004b. As of October 7, 2005:
http://www.gao.gov

———, *Military Personnel: Preliminary Observations on Recruiting and Retention Issues Within the U.S. Armed Forces*, GAO-05-419T, March 16, 2005b, pp. 13–16. As of October 7, 2005:
http://www.gao.gov

———, *The Army's Future Combat System's Features, Risks, and Alternatives*, GAO-04-635T, April 1, 2004a. As of October 7, 2005:
http://www.gao.gov

U.S. Joint Staff, "Joint Capabilities Integration Development System (JCIDS)," CJCSI 3170.01, April 15, 2001. As of December 19, 2005:
http://www.dtic.mil/doctrine/jel/cjcsd/cjcsi/3170_01b.pdf

U.S. Naval Institute, "Nuclear-Capable Weapons Dealt to Iran, China," *U.S. Naval Institute Periscope Database*, February 22, 2005. As of December 19, 2005:
http://www.periscope.ucg.com/special/special-200502221108.shtml

———, "F-22A Rapor Fighter," U.S. Naval Institute Periscope Database, n.d. As of February 22, 2007 (available by subscription only):
http://www.militaryperiscope.com/weapons.aircraft/fighter/w0003150.html

U.S. War Department, *Annual Report of the Secretary of War*, Chief of Staff testimony, Washington, D.C., 1920, pp. 244–245.

———, *Report of the Secretary of War to the President*, statement of Assistant Secretary of War Hurley, Washington, D.C., 1929.

Weigley, Russell F., "The Interwar Army, 1919–1941," in Kenneth J. Hagan and William R. Roberts, eds., *Against All Enemies: Interpretations of American Military History from Colonial Times to the Present*, Westport, Conn.: Greenwood Press, 1986.

———, *History of the U.S. Army*, Bloomington, Ind.: University Press, 1984.

"What Is the Real Health of the Defense Industrial Base?" *Manufacturing and Technology News*, February 22, 2005, Vol. 12, No. 4. As of December 19, 2005:
http://www.manufacturingnews.com/news/05/0222/art1.html